THE
NEMESIS
AFFAIR

Also by David M. Raup

Handbook of Paleontological Techniques
  (edited with Bernhard Kummel)

Principles of Paleontology
  (with Steven M. Stanley)

The Evolution of Complex and Higher Organisms
  (edited with D. H. Milne, John Billingham,
  Kevin Padian, and Karl Niklas)

Patterns and Processes in the History of Life
  (edited with David Jablonski)

Extinction: Bad Genes or Bad Luck?

THE NEMESIS AFFAIR

*A Story of the
Death of Dinosaurs
and the Ways of Science*

David M. Raup

Revised and Expanded

W·W·Norton & Company·New York·London

First published as a Norton paperback 1987;
revised and expanded paperback edition 1999.
Printed in the United States of America.
The text of this book is composed in Times Roman, with
display type set in Firmin Didot. Composition and
manufacturing by Allentown Digital Services Division of
R.R. Donnelley & Sons Company
Book design by B. L. Klein

Library of Congress Cataloging-in-Publication Data
Raup, David M.
The nemesis affair.
Includes index.
1. Earth—Origin. 2. Science—History. 3. Dinosaurs.
I. Title.
QB631.R38   1986   560   86-761

ISBN 0-393-31918-0

W. W. Norton & Company, Inc., 500 Fifth Avenue, New York, N.Y. 10110
www.wwnorton.com

W. W. Norton & Company Ltd., 10 Coptic Street, London WC1A 1PU

1 2 3 4 5 6 7 8 9 0

To all the mavericks who keep challenging the conventional wisdom. They think up new ideas just for the fun of it or out of innate contrariness. Without them, we would be wrong even more of the time.

Cartoon upper left by Slug Signorino, from *Chicago Reader*

Contents

List of Illustrations

Acknowledgments

There have been times over the past three years when my confidence in the extinction research has faltered. The statistical analyses Jack Sepkoski and I used to claim periodicity of extinction could easily have turned out to be wrong and thus we would have put a lot of other scientists and many journalists to work for nothing. It was at one of these low points that I conceived the idea of writing this book—as a fail-safe strategy. If the whole research program came tumbling down, I could describe its history as scientific discovery gone wrong. The more I thought about it, the more I was intrigued by the idea of publishing a chronicle of scientific failure. And I am indebted to Nancy S. Philippi for providing a sounding board and for keeping the idea alive through some difficult times. As it turned out, I was too impatient to wait for the outcome. And the next few years will tell us all whether this has been an account of success or failure.

I am most grateful to a number of friends and colleagues who helped with the manuscript. In particular, I would like to thank Susan Alexander, Joan Chandler, Stephen Jay Gould, Anne Hornickel, David Jablonski, Daniel McShea, William Provine, Mickey and Marian Raup, and David Walsten. Glenda York was ever-helpful with logistics and Mary Wall drafted illustrations on a tight time schedule. And Lorna Gonzales insulted me sufficiently to keep the project going.

Special acknowledgment is due those scientists who played critical roles in the development of the Nemesis Affair but have been inadvertently left out of my narrative. Two of these are Victor Clube and Dale Russell: mavericks who have pressed unconventional explanations of mass extinction and have not received enough credit.

My editor at Norton, Edwin Barber, is everything that an editor should be and I am grateful for his help and wisdom. Also, I must acknowledge my parents, Hugh and Lucy Raup, for many years of undaunted support and, above all, for lessons in how to be a constructive maverick. Finally, the events described in this book owe much to innumerable research colleagues, in Chicago and around the world. Of these, I owe special thanks to Jack Sepkoski with whom I have been immersed in extinction problems for several years. No matter how the science comes out, this association has been most rewarding.

Introduction 1999

For as long as humankind has scanned the night sky, we have known that the Earth is not alone. Only recently, however, have we come to appreciate the number and variety of objects out there and their potential for colliding with Earth. As late as the 1950s, we saw no danger in the many comets and asteroids that share our solar system. One hole in the Earth—Meteor Crater in Arizona—was grudgingly accepted as the impact of a small asteroid, but surely only a lone rogue. Small, rare, and not very important. That bit of guesswork— hopefulness?—has turned out to be dead wrong.

Now, geologists and astronomers have hard evidence of numerous impact craters on Earth, many far larger than Meteor Crater. And sightings of Near Earth Objects (NEOs) have produced a catalog currently listing 283 asteroids with orbits crossing the orbit of Earth (now available on the Internet at http://impact.arc.nasa.gov). The crater and NEO data have made possible reasonably good estimates of the likelihood of a catastrophic impact on human time scales. The risk, it seems, is very low but it is not negligible. Astronomers Clark Chapman and David Morrison, in their authoritative book *Cosmic Catastrophes* (1989), have calculated that the lifetime risk of being killed in the aftermath of an extraterrestrial impact is about the same as dying by accidental electrocution and substantially greater than being killed in an airplane crash.

The obvious box office appeal of disaster by cosmic collision has been exploited in both fiction and nonfiction (with some of the nonfiction being pretty close to fiction), famously last season with the hit film *Deep Impact*. I haven't seen the film because the small island in Lake Michigan where I live has no movie house, but the production is being lauded by the scientific community for its remarkable accuracy and realism—pretty rare for Hollywood's attempts at science!

Whatever the film's virtues, a large comet or asteroid impact almost certainly did play a decisive role in the extinction of the dinosaurs 65 million years ago. The ecological "space" provided by this mass extinction probably gave our mammalian ancestors the boost they needed to evolve and diversify, with one result being us! If this is true, the human species owes its existence to a large rock falling out of the sky. So much for environmental stability always being good for the evolution of life.

As a paleontologist, I have been immersed for two decades in the implications of large-body impact for the history of life. Have collisions with asteroids and comets been frequent enough and large enough to have significantly influenced the overall course of evolution? Can individual craters or other evidence of impact be positively linked to extinction events? Are large impacts predictable, in the sense of being uniformly spaced in time—like sunspots or El Niño—but on geologic rather than human time scales? This last question has involved me particularly because a colleague (Jack Sepkoski) and I proposed in 1984 that past pulses of extinction tend statistically to occur on a 26-million-year cycle: a uniform periodicity. Much of this book is devoted to the periodicity proposal and the controversies that have surrounded it.

More broadly, however, this book is a chronicle of five years of incredibly rapid progress and change in our thinking about the history of comet and asteroid impacts and their biological consequences. Because of my involvement in several aspects—especially the 26-million-year periodicity—it is a "view from the trenches." In the trenches were some strange companions: specialists in scientific disciplines that rarely talk to one another. Astronomers, physicists, geologists, geochemists, and paleontologists suddenly found common ground and a need to communicate. Through this, I learned a lot about the way science works, warts and all. And the popular press was fully involved because anything about dinosaurs has appeal and because predictions of cosmic collisions in our own future are bound to attract attention. Journalists even joined us occasionally in the trenches. They gave me far more insight into the problems of the public understanding of science. So, I tried in the text that follows to describe (and sometimes interpret) more than just the discoveries that have led to our present understanding of the influences of natural space junk in our solar system. Even though some of the present-day "facts" may fall by the wayside in years to come, the ways of science won't, for they are universal.

Washington Island, Wisconsin

THE
NEMESIS
AFFAIR

1: *Death Star*

THIS is the story of an emerging scientific theory about the extinction of dinosaurs and other prehistoric life forms. But it is also an account of the way science works, as seen by a participant. Five years ago, most paleontologists were rather complacent in their ignorance about the causes of extinction. To be sure, there were many theories, but I don't think most people thought the complexities of the great mass dyings would ever be unraveled. As a paleontologist, I shared the conventional view: extinction was a fascinating problem but not one that would yield to simple solutions.

All this has now changed, and the once rather old-fashioned science of paleontology finds itself in a maelstrom of excitement and controversy. Astrophysicists, atmospheric scientists, geochemists, geophysicists, and statisticians are all contributing to the extinction problem. And the general public is taking part through television talk shows, magazine cover stories, newspaper editorials, and even the occasional mention in gossip columns.

The story of the past five years in extinction research, and especially the role of the "Death Star" called Nemesis, is interesting in its own right and I will try to describe it in a reasonably orderly way. But my stronger reason for writing

this book is to say something about the way scientific research works and how it interacts with contemporary society. Science is not the pure, isolated endeavor that is usually
depicted. It is rarely a simple process of posing hypotheses,
devising experimental tests, and waiting for Yes or No answers. Although the answers are occasionally simple, getting
them published and accepted by the scientific community is
not. And scientists are victims of the same emotions and
belief systems as other people.

NEMESIS: THE THEORY

*Nemesis is one of several names given to our Sun's small
companion star. This little star is now about about two light
years away and moving away. But in another few million
years, it will turn and head back toward the Earth. The
inward trip will take another dozen or so million years, and
before the orbit is complete, Nemesis will have passed close to
the Sun. Close enough, in fact, to pass through the Oort Cloud,
an envelope of billions of comets that go around the Sun in
their own orbits beyond the outer planets. As Nemesis passes
through the Oort Cloud, its own gravitational force will deflect
some of the comet orbits in random ways. Most of them will
be thrown out of the Solar System but some will be sent in
toward us. As a result, one or more of the errant comets will
collide with Earth. And we know from the geological records
of Earth history that such collisions can be devastating. One
incident killed the dinosaurs and another got the last of the
crab-like creatures called trilobites. Many of the major biological crises of our past, the mass extinctions, were evidently
caused by the environmental shock of what is known in the
trade as "large-body impact." And because the Nemesis orbit
has a fixed period of 26 million years, the biological catastrophes come every 26 million years. A great big clock in the
sky is controlling biological destinies on Earth.*

The Nemesis story is now familiar, at least in outline, to readers of *Time, Newsweek,* any number of popular science magazines, and most major newspapers. The *New York Times* has published at least three strong editorials on the subject, and *Time* has had a long cover story on Nemesis. The science in these writings is good enough as far as it goes, but the course of research is far more complex.

In spite of the hype, there are a few inescapable facts. First, nobody has ever seen Nemesis and there is no direct, observational evidence that our Sun has a companion star at all. The Oort Cloud of comets has never been seen. And the demise of the dinosaurs by comet impact is debatable. The 26-million-year periodicity in mass extinctions may or may not be real—it is a matter of fairly abstruse statistical inference with rather messy data. In short we do not know whether any of the Nemesis story is true. But neither is it a Kiplingesque Just-So Story, because there is some evidence for all parts of the story. It is a novel theory in the process of testing.

Heated controversy over the Nemesis idea is rampant in the scientific community. To some observers, the "discovery" of the Death Star heralds a completely new way of looking at the Earth and its life, a scientific revolution. To others, it is science gone mad. Yet the idea has been with us only since late 1983 and the research that might move us from theory to fact has only begun.

CONNECTIONS

One newspaper reporter described the Nemesis story as having everything but sex and the Royal Family. It has even been suggested that the Death Star's name should be changed to Diana so that everything of human interest could be included.

Let me run through some of the connections the Nemesis

story has with other things. First, of course, it says or claims
to say something about the death of the dinosaurs, which is
something almost all children and many adults have won-
dered about for generations. We have heard all sorts of explanations for the extinction, from constipation to sterility, and
Kipling himself would not have dared publish some of the
wilder ideas that have been proposed. But the dinosaurs did
die, and the causes of that event may have something to say
about global habitability in general.

Are we living on a safe planet or should we have chosen
a better one? Did the dinosaurs do something wrong? However they died, one thing is fairly clear: their absence following the mass extinction 65 million years ago provided the
space for our mammalian ancestors to evolve and diversify.
Humans are probably here now because (among many other
factors) of the death of the dinosaurs. So, any credible and
testable explanation for the dinosaur extinction is both interesting and important.

And this leads to more immediate questions of global
habitability. What are the chances of comets or asteroids
hitting the Earth now? In 1908, an extremely small comet
fragment exploded over the remote area of Tunguska in
Siberia and knocked down about 6,000 square miles of forest.
It is not entirely clear whether this was a rare occurrence, a
rogue event, or whether it is what should be expected on
scales of human lifespans, just as we expect hurricanes and
tidal waves. If comet impacts are really concentrated at intervals of 26 million years, as the Nemesis hypothesis suggests,
do we have relative immunity to such crises most of the time?
If so, Nemesis is comfortably distant from the Oort Cloud,
at least for the next dozen million years.

This leads to another set of problems. If a Tunguska-like
impact does occur in our time, what is the possibility that it
will be misinterpreted? Could it be confused with a missile
launched by a foreign antagonist? As it happens, the U.S.

Defense Department and its counterpart in the Soviet Union have already thought of this and have set up procedures— effective, I hope—to guard against such confusion. In so doing, the notion of actually controlling the trajectories of incoming comets or asteroids has also come to the fore. When a comet or asteroid is a substantial distance from the Earth, its orbit could be altered with relative ease by shooting something at it. And this strategy could be used either defensively, by altering a collision course, or offensively, by changing the trajectory of the incoming body to aim it at a specific point on Earth.

A particularly curious aspect of the Nemesis story is its connection with the nuclear-winter scenario we have heard so much about in the past couple of years. Shortly after we learned of hard evidence for a large-body impact coinciding with the dinosaur extinction 65 million years ago, several geophysicists and atmospheric scientists did experiments and calculations to estimate the environmental effects of a collision with a large hunk of rock or ice. The impacting body has been estimated at about six miles in diameter. While all agreed that the impact of a body so large would have profound effects on the Earth's surface environment, sophisticated computer simulations were necessary to estimate more or less exact effects. The most widely accepted (but not necessarily correct) scenario to come from this research calls for our atmosphere to be so choked with fine debris and water vapor that the entire Earth becomes dark. This in turn would prevent photosynthesis in green plants and cause the demise of active animals dependent on plants for food. The plants themselves could survive the period of darkness and cold if it were short enough. The interesting point here is that the experiments and computer modeling that went into the dinosaur-extinction problem were soon picked up by Carl Sagan and others and applied to the environmental effects of a major thermonuclear war. And nuclear winter, in its American version, was born.

There is even a connection with the ongoing search for extraterrestrial intelligence—the SETI (Search for Extraterrestrial Intelligence) program of NASA and its counterparts in other countries. Part of the search strategy for life elsewhere in the universe has been to select those star systems most similar to our own, on the theory that what happened here is most likely to have happened in similar situations elsewhere. Because about three-quarters of the stars in our galaxy are double or multiple stars, and because our sun has been assumed to be a single (lone) star, it has been natural to concentrate the SETI search on single stars elsewhere. It has been assumed that life as we know it would be unlikely to evolve in an unstable system wherein two or more suns are orbiting around a common center. This is probably true for double stars where the suns are about the same size. But the possibility of Nemesis as a small second star raises the possibility that the evolution of complex life may thrive on (or even require) the adversity of this kind of double-star system.

This is just a glimpse of the wonderful and sometimes weird connections the Nemesis story has with other aspects of our natural world and of human affairs. The Nemesis theory may turn out to be a major step forward in our understanding of the natural world or an embarrassing period of near-insanity in scholarship.

Although this is not an autobiographical book, I should declare myself and my background. I was born in Boston in the mid-1930s to an academic family. Both my parents are botanists and my father was on the Harvard faculty for many years before retiring in 1967. I was trained in geology and paleontology at Chicago and Harvard and, after a brief fling with the petroleum industry, have been an academic through most of my career. For the past several years, I have been at the University of Chicago, an institution which, to my biased view, is the strongest and most vibrant intellectual center in

America. I currently hold faculty appointments at Chicago in geophysics, evolutionary biology, and the conceptual foundations of science.

My research over the years has been varied, and successful in the sense that I have accumulated far more recognition than I expected. It has been, and continues to be, fun and rewarding. I have worked almost exclusively in paleontology, but not in the kind of paleontology most people hear about. No great field expeditions, no work on the fossils of early humans, and no yen to unearth and describe new forms of past life. I have never described or named a new fossil species, a fact that caused some amusement when I was elected president of the Paleontological Society some years ago.

Fascinated by some of the general problems around the fringes of the conventional practice of paleontology, I have spent a lot of time doing highly theoretical, mathematical studies of the evolution of life. This got me into computers as early as the late 1950s, with FORTRAN programs to draw pictures of ideal snails and other molluscs, and with simulation models of the evolutionary process. This work made it feasible to ask questions like "What is the range of all possible snail shells, including those that have never appeared on Earth?" or "What would evolutionary patterns have looked like if Darwin had been wrong?"

My research has been a little jarring to some of my more conventional, specimen-oriented colleagues, but I have gotten away with it—at least until Nemesis came along. To me, fossils are fun to a point, but so are crystals and lots of other natural objects. The really interesting problems involve the far-out questions of theory. The big picture. This is risky because the "answers" are more often wrong than right.

A BRIEF CHRONOLOGY OF THE NEMESIS AFFAIR

The companion star Nemesis was first named in an article in the April 19, 1984, issue of the British journal *Nature,* but we must go back to 1980 to find the important underpinnings of the story. The summary will inevitably be biased in favor of the events I witnessed.

June, 1980: The journal *Science* published an article entitled "Extraterrestrial Cause for the Cretaceous-Tertiary Extinction," by Luis Alvarez, Walter Alvarez, Frank Asaro, and Helen Michel (all of the University of California at Berkeley). This was the first general announcement of the extraterrestrial-impact hypothesis of dinosaur extinctions based on a finding of anomalously high concentrations of the element iridium at the geological boundary between the Cretaceous and Tertiary periods (the so-called *K-T boundary*).

1980 to the present: Field and laboratory research by several groups in various countries continued to explore the Alvarez hypothesis of large-body impact as a cause of mass extinction. Many more *iridium anomalies* were found worldwide, as well as evidence of other extraterrestrial signatures at the K-T boundary. Generally strong opposition to the idea developed among rank-and-file paleontologists and geologists, although this is undergoing a gradual shift toward the impact hypothesis.

July, 1981: NASA's Ames Research Center convened the first of three workshops on the evolution of advanced life, extraterrestrial influences on evolution, and advanced life in the universe. I chaired the workshops, and the group included several people who were to become prominent in the Nemesis debates.

October, 1981: A conference, "Geological Implications of Impacts of Large Asteroids and Comets on the Earth," was held at Snowbird, the ski resort in Utah. This meeting, soon to become known as the Snowbird Conference, was sponsored by the Lunar and Planetary Institute and the National Academy of Sciences and attracted about 120 people from many disciplines for discussion and debate of the Alvarez hypothesis. The debates were so intense that the dramatic scenery and superb weather at Snowbird were all but ignored by the participants.

Spring, 1983: My colleague J. John (Jack) Sepkoski, Jr., and I started "number crunching" the computerized version of Sepkoski's *Compendium of Fossil Marine Families* that had been published in 1982. We thought we saw the cyclical or periodic behavior of extinctions reported by Alfred G. Fischer and Michael A. Arthur in 1977, which had been ignored or vehemently rejected by almost all of us.

May, 1983: I attended a conference in Berlin, "Patterns of Change in Earth Evolution" and presented our tentative results on periodic extinction. The presentation was low key. I kept it out of the conference report and tried to keep it out of the press, with some success.

August, 1983: Our periodicity story had become a bit firmer and Jack Sepkoski made a short presentation at a symposium, "The Dynamics of Extinction," in Flagstaff, Arizona, and suggested that some unknown extraterrestrial driving force was responsible for periodic extinction.

September, 1983: Accounts of Sepkoski's Flagstaff presentation appeared in *Science, Science News,* and the *Los Angeles Times,* thus providing non-geologists and the general public with their first summary of the idea. This sparked the flurry of interest among astronomers and astrophysicists that was to crystallize in the April 19, 1984, issue of *Nature.*

October, 1983: Sepkoski and I submitted a short paper reporting our statistical results, under the title "Periodicity of Extinctions in the Geologic Past," to the *Proceedings of the National Academy of Sciences (PNAS).*

February, 1984: Our *PNAS* paper was published: five pages of statistical analysis arguing for the 26-million-year periodicity and concluding that the periodicity is probably driven by solar system or galactic forces. We suggested that passage of the Solar System through the spiral arms of the Milky Way galaxy might be the culprit.

April 19, 1984: Nature published five papers giving astrophysical interpretations of the 26-million-year periodicity: Michael R. Rampino and Richard B. Stothers, and Richard D. Schwartz and Philip B. James, on the Solar System's vertical motion through the Galaxy; Marc Davis, Piet Hut, and Richard A. Muller, and Daniel P. Whitmire and Albert A. Jackson IV, proposing the companion star, dubbed "Nemesis" in the Davis *et al.* paper; and Alvarez and Muller on finding a similar periodicity in the ages of meteorite craters on Earth. The five papers were preceded by rather negative editorial commentaries by John Maddox (*Nature*'s chief editor) and Anthony Hallam (professor of geology at Birmingham).

January, 1985: Science Digest used an account of Nemesis as the first of its "Stories of the Year" (1984) for astronomy and physics.

January, 1985: Daniel P. Whitmire and John J. Matese published a paper in *Nature* proposing that the 26-million-year periodicity was actually due to the motion of an unrecognized Planet X in the neighborhood of Neptune.

January, 1985: In Tucson, a special session of the annual meeting of the American Astronomical Society was devoted to debating the Nemesis hypothesis. A number of brickbats were thrown at each of the prevailing hypotheses. Sepkoski's and my statistical analysis of the extinction data came under fire from Scott Tremaine of MIT.

March, 1985: Richard Kerr published a Research News column in *Science* (based on the Tucson meeting) under the title "Periodic Extinctions and Impacts Challenged."

March, 1985: My paper claiming a 30-million-year periodicity in reversals of the Earth's magnetic field was published in *Nature*.

April 2, 1985: The *New York Times* ran its first editorial on the subject ("Miscasting the Dinosaur's Horoscope"), which ended with the now-famous statement: "Astronomers should leave to astrologers the task of seeking the cause of earthly events in the stars."

May, 1985: Time published a cover story on periodicity and Nemesis.

June, 1985: The first state-of-the-art dinosaur book for children describing the Nemesis story was published: *The Dinosaurs and the Dark Star,* by Robin Bates and Cheryl Simon. Also, at about this time, Shriekback, the British rock group, recorded a song based on Nemesis. And catastrophic impact from space became a more common ingredient of science-fiction stories.

June, 1985: Antoni Hoffman, a paleontologist at Columbia, published a paper in *Nature* lambasting the Sepkoski–Raup statistical analysis of the extinction record. This was introduced by a John Maddox editorial strongly endorsing the Hoffman piece.

July, 1985: The *New York Times* ran its second editorial ("Nemesis of Nemesis"), supporting the Hoffman analysis.

October, 1985: The report of massive amounts of soot in clays at the K-T boundary was published in *Science* by Wendy S. Wolbach, Roy S. Lewis, and Edward Anders. The soot was interpreted to be the result of massive wildfires touched off by comet impact.

October, 1985: A challenge by Timothy M. Lutz to my periodic magnetic reversal paper was published in *Nature*. The paper was introduced by my own News and Comment piece congratulating Lutz but defending periodic extinction.

October, 1985: Stephen Jay Gould published a strongly worded essay in *Discover* criticizing Hoffman, *Nature*, and the *New York Times*.

2: Catastrophism and Earth History

CUVIER VERSUS LYELL

MUCH of the debate over the Nemesis theory stems from the implications of the somewhat vague term "catastrophism." In many fields of the natural sciences, but especially in geology, catastrophism is a highly loaded word. Even an idle mention of the possibility that something in the history of the Earth could be called catastrophic can produce showers of denial and abuse from many geologists.

What is a catastrophe? It is certainly something sudden and, I think, something not predicted in advance. It is big! That is, big in comparison to what is normal or expected. There is also an element of misfortune. One does not describe something as catastrophic unless somebody or something loses. In the natural world, large earthquakes, floods, and volcanic eruptions all qualify.

For the geologist, there has long been a question of whether catastrophic events accomplish more change in the long run than the sum of everyday calmer, background processes. This was debated long and hard in the mid-nineteenth century, and the results of that debate form a vital foundation for any discussion of rocks that fall from the sky and kill dinosaurs, whether this happens on a set time schedule or erratically. The nineteenth-century debate is sufficiently important and interesting to trace in part.

One of the greatest naturalists of all time was the French anatomist-paleontologist, Baron Georges Cuvier (1769–1832). Working mainly in the lushly fossiliferous deposits of the Paris Basin, he was a pioneer in working out geologic sequences and the history of life. Much of Cuvier's research is as good today as it was 150 years ago. It is important to note that Baron Cuvier (and his many counterparts elsewhere in Europe) were working decades before the 1859 publication of Charles Darwin's *Origin of Species.*

From his observations in the Paris Basin, Cuvier developed strong ideas about Earth processes and the course of biological history. He saw both as being highly punctuated by sudden change. Organisms would appear in the fossil record, remain unchanged for long periods, and then disappear suddenly. And he interpreted this style of change to be one dominated by occasional catastrophes. He wrote in 1817, for example:

> These repeated [advances] and retreats of the sea have neither been slow nor gradual; most of the catastrophes which have occasioned them have been sudden; and this is easily proved, especially with regard to the last of them, the traces of which are most conspicuous. . . . Life, therefore, has often been disturbed on this Earth by terrible events—calamities which, at their commencement, have perhaps moved and overturned to a great depth the entire outer crust of the globe . . . numberless living things have been the victims of these catastrophes . . . Their races have even become extinct . . . (translated from French)

This view of the history of the Earth and of life was challenged by the great British geologist Sir Charles Lyell

(1797–1875). He interpreted the same sequences in very different ways. For example, he wrote in the first edition of his famous textbook on geology (1833) the following rather purple prose:

> We hear of sudden and violent revolutions of the globe, of the instantaneous elevation of mountain chains, of paroxysms of volcanic energy . . . We are also told of general catastrophes and a succession of deluges, of the alternation of periods of repose and disorder, of the refrigeration of the globe, of the sudden annihilation of whole races of animals and plants, and other hypotheses, in which we see the ancient spirit of speculation revived, and a desire manifested to cut, rather than patiently to untie, the Gordian knot.

Lyell disagreed with Cuvier and he was pretty nasty about it. The phrase "ancient spirit of speculation revived" is definitely not meant as a compliment. And his bit about patiently untying the Gordian knot says that only by honesty and hard work are geologists going to get anywhere. Catastrophism is seen as a quick fix.

As many readers will already have realized, the debate and argument about Nemesis is a revival of the Lyell–Cuvier argument.

Lyell went on in his textbook to hammer his point home and even give it some moral overtones. He wrote:

> In our attempt to unravel these difficult questions, we shall adopt a different course, restricting ourselves to the known or possible operations of existing causes; feeling assured that we have not yet exhausted the resources which the study of the present course of nature may provide, and therefore, that we are not authorized in the infancy of our science, to recur to extraordinary agents. We shall adhere to this plan . . . because . . . history informs us that this method has always put geologists on the road that leads to truth.

Lyell sounds a little more like a lawyer (his original training), a preacher, or a politician than a scientist. He is clearly making almost a moral issue of catastrophe.

The Lyell–Cuvier debate raged for some years, with students and disciples active on both sides. That the Lyellian forces won hands down has profoundly affected thinking in geology and a number of cognate fields to the present day. It is difficult to exaggerate the power of Lyell's victory and the depth of Cuvier's defeat. I had a strange opportunity to see some of this from the French side in the summer of 1985. A French colleague and I were talking over dinner about the present state of French evolutionary biology and paleontology. My friend commented that French scientists of his acquaintance were very hesitant to propose new theories or break new ground in their fields. I asked him why, thinking of all sorts of pop sociological reasons, from Mitterrand to wine. His answer shocked me. He said that his colleagues had never gotten over the Cuvier debacle and wanted to avoid the same thing happening again.

Anyway, because of the now-pervasive Lyellian stance, it has always been more acceptable in geology to avoid the temptation "to recur to extraordinary agents" in any interpretation of Earth history. This is not to say that the Lyellian paradigm has not served us well—because it has. Many geological phenomena are clearly best understood and most correctly explained in terms of well-known and regular processes that can be observed, measured, and even experimented with in the modern world. The present is indeed the key to the past.

Also, and perhaps more important in some ways, the Lyellian dogma has made it possible to maintain a clear separation between science on the one hand and the lunatic fringe and religion, on the other. I don't mean to equate the last two, but both are commonly seen as nuisances, at the very least, to organized science.

The Lyellian paradigm has had some interesting effects. It is taught in the classroom under the name "uniformitarianism" and represents a catechism that all geology students learn and learn well, even though there is always some confusion about just what uniformitarianism means. A classic example of some of the problems this has caused is the case of J. Harlen Bretz and the Channeled Scablands.

The Scablands is an area in eastern Washington (south of Spokane) where deep channels have been eroded into thick glacial soil deposits and into the underlying volcanic rock. The channel bottoms are filled with coarse gravel carried in from well outside the area. "Doc" Bretz did extensive field work on the Scabland landscape in the years following World War I. He interpreted much of what he saw to be the result of a gigantic, catastrophic flood caused by glacial meltwaters. To Bretz, the depth of the channels, the erosion and scouring of the volcanic rock, and the gravel filling all pointed to sudden flooding by huge amounts of water, rather than the slower, calmer action of ordinary stream and river flow. Bretz was roundly denounced for his calling on an "extraordinary agent" and might have been drummed out of the science had he not held a tenured faculty position at Chicago. His flood postulation, reminiscent of the Noachian Deluge, was totally unacceptable to the geological community. To be sure, Bretz's case was somewhat weak, especially as he had no credible source of a sudden deluge.

Scientific vindication finally came for Bretz, however, after many discouraging years. Two new kinds of information came to light. First, geologists working independently in western Montana found evidence of what had been a large glacial lake. Because this lake was probably held in place by ice dams, it provided a source for floodwaters. Second, aerial photography (and later satellite imagery) developed to the point where landscape features could be seen that were invisible from the ground because of their scale. The photographs showed ridges on several of the channel floors that could only be giant ripples of the type found on smaller scales on the bottoms of most fast-moving streams. The new evidence was decisive and Bretz's deluge became credible in spite of the Lyellian resistance to the idea.

Fortunately, Bretz lived to enjoy his victory, although not until 1976, when, in Bretz's 96th year, the Geological Society of America gave him its highest award, the Penrose Medal. The Bretz episode was a good lesson for all in science although it did not have much fundamental effect on textbooks or indeed on general practice in the Earth sciences.

METEORITES

Meteorites, literally "rocks that fall out of the sky," certainly qualify as examples of Lyell's "extraordinary agents." The history of geological views about meteorites is interesting not only from the Lyellian viewpoint but also because meteorites play an important role in the Nemesis Affair.

"Meteorite" is the term given to any rock found on Earth that is of extraterrestrial origin, regardless of where in the Solar System the rock originated or whether it began life as an asteroid or comet. Meteorites tend to be quite distinctive and readily identifiable; some have actually been seen to hit the Earth. Most of them are thought to be fragments formed by asteroids colliding with each other in the zone between the orbits of Mars and Jupiter. The collisions deflect the fragments, throwing some into Earth-crossing orbits. A fraction of these are large enough to survive the trip through the Earth's atmosphere.

What about comets? The general term meteorite is applied to any rock of extraterrestrial origin because there is some uncertainty as to whether they are comets or asteroids. Comets undoubtedly hit the Earth and make craters, but none has ever been positively identified (except perhaps at Tunguska).

Our inexperience of comet collision ties in to our ignorance of the composition of comets themselves. The consensus is that they are "dirty snowballs"—mainly ice studded with rock fragments. But we have not had specimens to analyze. To make matters worse, there is some possibility that the asteroid population of the Solar System is renewed over long periods of time by addition of degraded comets from the Oort Cloud.

Meteorites themselves have been known for a long time but only recently have geologists granted that meteorites large enough to form craters regularly bombard the Earth. As late as 1964, the eminent Harvard professor Kirtley Mather wrote that only five or six impact craters exist on the Earth, the best known being Meteor Crater in Arizona. For many years, the craters on the Moon were thought to be volcanic rather than impact structures.

These notions changed totally with the development of accurate physical and chemical ways to identify meteorite craters. Also, satellite photography has made it possible to recognize craters too large to see from the ground or in conventional aerial photographs. Along with this has come geologists' much greater willingness to consider an impact origin for crater-like features on the Earth. At present, about 100 impact craters have been authenticated. They range in age up to about two billion years and in size up to 140 kilometers in diameter. There are undoubtedly many more such craters to be found and a larger number yet that have not survived the ravages of erosion. There is every reason to believe that the Earth's surface would be as pockmarked as the Moon were it not for erosion.

An interesting aspect of the history of meteorite study is that the uniformitarian doctrines of Lyell and his followers have been able to absorb the new facts and concepts without seriously changing the basic catechism. Meteorites fall and they make craters, often big ones. Once accepted, this pattern became part of the uniformitarian doctrine and was no longer considered catastrophic. The following statement by George Wetherill and Eugene Shoemaker in 1982 illustrates this:

> Although the physical encounter with the Earth of these objects can properly be termed "catastrophic," in terms of the magnitude of the effects they produce, they are at the same time "uniformitarian" in that they represent the extension of presently observed geologic processes to earlier geologic time.

One can argue, therefore, that catastrophism really means something that is unfamiliar. As soon as it becomes familiar, the awful label need no longer burden it. Scientists are adaptable people.

MAVERICKS AND RASH PROPOSALS

Throughout the past century, geological and paleontological thought has been dominated by Lyell's views. Cuvier is often cited for comic relief in introductory geology classes and occasionally as a weapon to put down an unconventional idea. All too often, a vaguely catastrophic proposal has been denied with a call to Baron Cuvier: it is wrong because Cuvier was wrong. Even more absurd have been the attempts to claim that Cuvier really didn't mean it. It is said that the English word "catastrophe" is a mistranslation of Cuvier's French, an idea that I find far-fetched because Cuvier's total context seems abundantly clear.

In spite of all the brainwashing, there have been occasional mavericks who have made proposals about extinction that are definitely catastrophic. I will describe a few that are especially germane to the Nemesis issue. They all have to do with mass extinction through extraterrestrial causes.

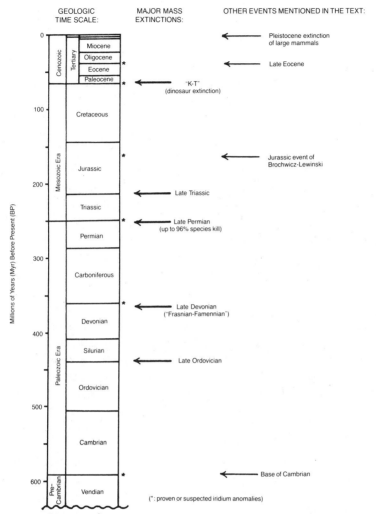

Geologic time scale for the past 600 million years of Earth history. The sequence of eras (Paleozoic, Mesozoic, Cenozoic) and periods (Cambrian, Ordovician, Silurian, and so on) is based primarily on fossils. The chronology is calibrated by isotopic dating to produce a time scale in millions of years. Several of the major extinction events are indicated by the arrows in the center column.

OTTO SCHINDEWOLF

Schindewolf was a professor of paleontology at Tübingen University in southern Germany from the end of World War II until his death in the spring of 1972. He was certainly the most respected scholar of the fossil record in Germany and perhaps in the world, widely known for his research on the great mass extinction at the end of the Permian period, 250 million years ago. Over the years, he did extensive field work in the Salt Range of Pakistan, one of the very few nearly complete fossil sequences across the boundary between Permian- and Triassic-age rocks. In 1962, he published a paper entitled "Neokatastrophismus?" Schindewolf's proposal was that the Permian mass extinction had been caused by a nearby exploding star, a supernova.

A star exploding close to the Earth would surely have disastrous effects on the global biota, but much would depend on how close the star was. Isaac Asimov has reported that if the supernova were only ten light years away, the visible and infrared radiation would produce a heat wave lasting many weeks and attendant climatic effects. Also, the atmosphere would be bombarded by high doses of X-rays and ultaviolet radiation. The biological effects are hard to gauge because we are still surprisingly ignorant about the consequences of radiation. Also, much of the radiation would be absorbed in the upper atmosphere. If the supernova were more than ten light years away, the effects would naturally be much less. There are historical records of sightings of very distant supernovae: in Europe in the years 1604 and 1572 and in China in 1054.

An important part of evaluating Schindewolf's proposal is its statistical credibility. What is the probability of at least one supernova occurring close enough to Earth to do substantial biological damage sometime in geologic history? It has been estimated that a supernova explosion within 100

light years may occur on average every 750 million years, and
this is a reasonable frequency for an event as unusual as the
Permian mass extinction. But if calculations mentioned
above are correct, 100 light years is not nearly close enough
to do the job. A supernova at the ten-light-year distance is
much less likely. Thus, postulating a supernova as an extinction explanation is really a long shot.

Far more important, Professor Schindewolf had no independent evidence at all for an exploding star in the Permian.
He made the proposal because he found the suddenness and
intensity of the mass extinction inexplicable in any other
way. He could think of no Earth-bound process or phenomenon that would explain what he saw in the fossil record. He
was, in a sense, acting in desperation. Had this idea come
from anyone of lesser experience with the Permian record
and paleontology in general, it probably would have been
ignored as naive. As it happened, the proposal caused very
little stir: a few papers published over the next few years
evaluated the biological effects of high-energy radiation, and
that is about all. Also, there was little any of us could do with
the proposal because of the lack of any geological trace of a
supernova in the Permian.

DIGBY MCLAREN

McLaren is a Canadian paleontologist and geologist of great
stature. He was Secretary General of the Canadian Geological Survey for several years and his excellence as a scholar
has been recognized in many ways, including the presiden-
cies of the Geological Society of America and the Paleontological Society and election as a Foreign Associate of the
National Academy of Sciences (USA). He now is on the
geological faculty of the University of Ottawa.

In his presidential address to the Paleontological Society
in 1970, McLaren talked extensively about his main love in

geology: the Devonian period and especially the mass extinction at the end of the Frasnian stage of the Devonian, about 365 million years ago. This extinction ranks as the fourth or fifth most intense of them all.

McLaren made the bold suggestion that the Frasnian extinction was an indirect result of an enormous meteorite impact. He was very careful, however, to present this only as one of the possible explanations. Like Schindewolf, he had no documentary evidence for the event and, also like Schindewolf, he was driven to the idea largely because he could think of no other plausible explanation for the suddenness and intensity of the event.

McLaren's proposal was more credible than Schindewolf's because large meteorite falls are known to be fairly common. By 1970, quite a number of large-impact craters on Earth had been verified and dated. But McLaren's presidential address and its subsequent publication caused even less commotion than Schindewolf's had about a decade earlier. This was in part because McLaren had no evidence, but I suggest that it was also because to geologists and paleontologists trained in the Lyellian catechism, the idea was unthinkable and therefore not worth arguing about.

It must have been sweet, then, for Digby McLaren to hear about Alvarez's discovery of the iridium anomalies at the Cretaceous-Tertiary boundary. And even sweeter when, in 1984, he was part of a team that found an iridium anomaly in Australia just where he expected it at the top of the Frasnian sequence.

HAROLD UREY

People generally listen to Nobel laureates. Harold Urey was a brilliant chemist who had broad interests and abilities in many areas of science. Among other things, he was a very respectable cosmochemist with more than a passing knowledge of geology. In the early 1950s, for example, he stimulated a group of chemists, geologists, paleontologists, and biologists at the University of Chicago in a project that was the first to use oxygen isotope ratios in fossils to deduce temperatures of the geologic past.

In 1973, Urey published a little paper in *Nature* arguing that several of the extinction events of the past 40 to 50 million years had been caused by impacts of large comets. Unlike his predecessors, Urey came armed with evidence. He looked at two kinds of historical records: the dates of extinction events in the Tertiary period and the ages of *tektites*. Tektites are small glassy blobs that are found occasionally in soils and rocks and generally agreed to be products of meteorite impact. Several episodes of tektite formation have been recognized and dated.

Perhaps because Urey did not have detailed knowledge of the fossil record of extinction, he used a proxy for the geologic dates of extinction events in the time span of the dated tektites. He used the radiometric dates of the boundaries between the subdivisions of the Tertiary (units like the Eocene, Oligocene, Miocene, and so on), being aware of the general fact that most boundaries in the geologic time scale are placed at significant extinction points.

He then built a statistical case to argue that the tektite ages and extinction times were too similar to be explained by chance. His interpretation: comet impacts as the cause of the extinctions.

Once again silence reigned. Urey's paper in *Nature* caused not a ripple despite his prominence and authority in science. I am not aware of a single paper citing the Urey work until the Alvarez paper in *Science* in 1980 on iridium and the K-T boundary event. Why?

Nature is very widely and carefully read. Perhaps a lot of people who read the paper did not like Urey's statistics and discarded it without fanfare or rebuttal. This is very common in science, because there is simply not enough time to fret over all the flawed papers that are published every month in scientific journals. And Urey's argument was a rather slim one.

The other possibility, of course, is that the Urey proposal was too far out to be heard against the backdrop of the Lyellian paradigm—that is, too incredible even to register, much less argue about. I have no recollection myself of having known about the paper at the time. If I had known about it, I would no doubt have found fault.

Urey's 1973 paper ended with the suggestion that tektites corresponding in age to the dinosaur extinctions at the end of the Cretaceous might someday be found. He thought it quite likely that comet impact caused that extinction as well. It is unfortunate that Urey did not live to read the paper senior-authored by another Nobel laureate, Luis Alvarez.

The examples of the catastrophic explanations for mass extinction put forward by Schindewolf, McLaren, and Urey are instructive in many ways. Most interesting, I think, is that each person was at the height of his career and influence when the proposals were made. Does this normally characterize maverick ideas? I am not enough of an historian of science to say. Were these cases of intellectual senility? Emphatically not! Perhaps crazy ideas continually sprout in the minds of some small fraction of scientists in all age groups, but it is only the leading people who have the personal confidence to try them out and the clout to get their papers accepted for publication.

In this context, here is one more example of a maverick approach to the problem of mass extinctions. This is from my own research in the mid–1970s and it has quite a different ending.

BOMBING AUSTRALIA

Throughout most of the 1970s, I was on the faculty of the University of Rochester. I was beginning to play with numerical analysis of extinction patterns in the fossil record. Jack Sepkoski's big *Compendium* of data had not been published —he was only beginning work on it as a graduate student under Stephen Jay Gould at Harvard—but I had a primitive data base on an IBM 7094. I don't think that I was aware of McLaren's 1970 meteorite/extinction idea, but I had stumbled on a series of papers written some years before by the Irish astronomer E. J. Öpik.

Öpik, a leading expert on comets, had thought for a long time that comet impact might cause biological devastation and extinction. His work was not widely known outside astronomy. I was attracted to the fact that Öpik was thinking not so much about global effects as about regional ones. He suggested that the damage of comet impact would be limited to what he called the "lethal area" around the impact point.

This interested me particularly because it is an old Cuvier idea. If, for example, all life in Australia and New Guinea were annihilated by a fairly localized disaster, this would produce the complete—and therefore global—extinction of those marsupial mammals that lived only in Australia and New Guinea. Because many plants and animals are naturally restricted to a single area or region, one does not need to call on a worldwide environmental shock to produce complete extinction of these "endemics." This provinciality might make it considerably easier to produce a mass extinction of the type we see in the fossil record.

How could this idea be tested, at least for feasibility? Our knowledge of the geographic distributions of fossil organisms is not nearly good enough to ask whether all the victims of a given extinction were restricted to one area or region. The next best thing is to simulate the effect using modern distributions of living organisms that are well known. So, I computerized the geographic distributions of a number of living animal groups: terrestrial birds, reptiles, amphibians, and mammals and fresh-water fish, as well as two marine groups: reef corals and echinoid echinoderms (sea urchins and sand dollars). This was fairly easy to do from the biogeographic literature.

I then "bombed" these distributions using a Monte Carlo simulation. This kind of simulation is so named because it employs computerized dice-throwing (a random-number generator) to mimic a complex and unpredictable natural process. The computer program chose random target points on the surface of the globe and assigned a lethal radius around each. Then, it was a simple matter of computerized bookkeeping to determine how many genera and families of animals could be declared "extinct" by virtue of their being totally contained within the lethal area. If the kill rates were in the range we normally designate as mass extinctions in the fossil record, then I would have a mechanism of mass extinction without global enironmental stress.

The results were somewhat unexpected. The simulations showed that a typical mass extinction could not be accomplished by complete killing in a region unless that region were extremely large—more than half the Earth's surface. I was a little frustrated by this because I had been thinking all along about the endemic marsupials in Australia, the endemic corals in the West Indies, and so on. I was so frustrated that I eventually removed the random-target part of the program and set up the computer so that I could aim the imaginary comets. This would allow me to annihilate all life in, say, Australia or the West Indies without fear that some parts of the region would escape. These catastrophic simulations were a source of much amusement for the graduate students and junior faculty at Rochester. They thought the whole project was crazy. But aiming the comets did not help very much. I still could not mimic a mass extinction by killing in small regions. There are a lot of endemic marsupials in Australia but they are only a small percentage of the global mammalian fauna.

When the research I have just described was complete, I naturally thought of publication. Even though the results were not quite what I had expected, it was a useful evaluation of the Öpik scenario for extinction in that I was able to put some mathematical limits on the effects of his scenario. *But I made no attempt to publish this research because I knew it would be laughed at, or worse.* So, I filed away all the data and computer programs and went on to other research projects.

A few years later, the atmosphere had changed. In October of 1981, I was invited to the Snowbird Conference on meteorite impacts and mass extinction. In need of a topic to present and write up for the conference proceedings, I have never had an easier job. I dusted off the old simulations and presented them under the title "Biogeographic Extinction: a Feasibility Test." It was not a great piece of work but it fit right in with Snowbird.

3: Dinosaurs and the Death of Species

ALL SPECIES ARE EXTINCT!

All species that have ever lived are, to a first approximation, dead. Scientists are fond of using phrases like "to a first approximation" or "to an order of magnitude" to express how similar two numbers are. Occasionally, these are just catch phrases to hide ignorance or to convince the listener that things are more precise than they really are. But usually these phrases have a definite meaning in a given context. "To a first approximation" means "almost" and "to an order of magnitude" means "within a factor of ten." Brooklyn and Queens are, to a first approximation, in the same place in our galaxy as a whole. On the other hand, Brooklyn and Queens are not in the same place by any approximation when viewed from Manhattan. All humans are, to an order of magnitude, the same height. So, it is in this vein that we can say that all species that have ever lived are dead. Extinction is the name of the game.

About one and a half million different species of living plants and animals have been recognized, described, and given latinized names. And by international procedures and rules, each of the names has been published, usually in recognized scientific journals. In spite of this rather incredible number, the known diversity of species is probably but a small fraction of the actual living diversity. Recent estimates, based on current rates of discovery, suggest that as many as 40 million species are alive today.

From these figures and from estimates of the average life-spans of species and the total duration of life on Earth, we can make some fairly good estimates of the number of species that have lived in the geologic past: the total progeny of species in the evolution of life. Thanks to some brilliant work by Leigh Van Valen at the University of Chicago about ten years ago, we have a good idea of how long species last, at least as statistical averages. The averages vary somewhat from one biologic group to another, but they all fall within a suprisingly narrow range: from about one to ten million years. Species durations are long in human terms but very short to a geologist who works with billions of years of Earth history.

Our oldest fossils are about 3,600 million years old, and the lush biology of multicellular organisms goes back about 600 million years. With a few calculations, we find that the species living today are only a minuscule percentage of the total number that ever lived, perhaps considerably less than 1 percent. There are arguments about some of the estimates we plug into the calculations, but all agree that most species are extinct.

The ubiquity of extinction was recognized early in the nineteenth century and was, in fact, vitally important in our learning to use the fossil record to classify time: that is, to build a chronology of Earth history. Because of extinction (and, of course, the origination of new species to replace the lost ones), assemblages of fossils change through time. The changes enabled the early paleontologists to piece together a sequence of events, just as an archaeologist uses changes in human cultures in a sequence of strata.

Extinctions are not uniformly distributed in geologic time. Some intervals, which we now call "extinction events" or "mass extinctions," have many more than the normal number of species going extinct. These intervals were used by the early nineteenth-century geologists to label points in time that could be recognized worldwide. Names given to the periods of time between the extinction events are still with us. Thus, it is not surprising that the major extinctions were, and still are, at major boundaries in the geologic time scale. It is no coincidence that the dinosaurs went extinct at, or very near, what we know as the boundary between the Cretaceous and Tertiary periods (the K-T boundary). This is also the boundary between two larger units: the Mesozoic and Cenozoic eras.

The greatest mass extinction of all time was in the Permian period, some 250 million years ago, at or near the Permian-Triassic boundary, which is also the era boundary between the Paleozoic and Mesozoic. It has been estimated that this event eliminated as many as 96 percent of species living in the oceans at that time: a nearly complete destruction of all life.

ORIGIN OF SPECIES

The foregoing discussion is mere bookkeeping. We know that a lot of species have originated on Earth and that most of them did not survive. But what about the processes responsible for this turnover?

Biologists have long been fascinated by the origin of species, but there are really two kinds of species origination. One is the "origin of species" that Darwin talked about: simple change in a single evolutionary line over time by natural selection. If this change is substantial enough, the descendant is a new species. The ancestral form does not go extinct in the sense of death, but rather it is changed or transformed into another species. The number of species coexisting at an instant in time stays the same.

The other kind of species origination is when a lineage of a species branches or buds to form another, coexisting species. More nearly equivalent to birth in a genealogical sense, this is the process that causes species populations to multiply and thus to provide the main fodder for extinction. To keep the terminology straight, the Darwinian kind of species transformation is called "phyletic transformation" and branching is called "speciation."

The branching process of speciation has been the subject of an enormous amount of research in evolutionary biology in recent years, deeply examined in international symposia and a number of important textbooks. Although the process still has mysterious aspects, we know a lot about it.

The number of species extinctions has probably been about the same ("to a first approximation") as the number of speciation events. If speciation were more frequent than extinction over hundreds of millions of years, the total population of species would grow out of proportion and there would be standing room only. By the same token, if extinction rates exceeded origination rates over long spans of geologic time, all life would go extinct.

So, we have two processes—origination and extinction—that are as different as birth and death but that, over the long run, have remained in reasonable balance.

THE EXTINCTION PROCESS

In marked contrast to speciation, biologists have learned almost nothing about the mechanisms of extinction. As we have seen, both processes are continuous (geologically, at least) and occur at about the same rate. The indexes of major textbooks on evolutionary theory, however, as well as those of scholarly treatises, hardly mention extinction. It seems that there has not been much interest in the subject.

When one does find discussion of extinction, it is usually dominated by platitudes and tautologies. "Species go extinct when the size of the breeding population approaches zero." Or: "Species die out because they can no longer cope." One does not have to know much about the subject to say things like this.

I can think of several reasons for the traditional lack of interest in, and ignorance of, extinction. One is that the problem may seem too simple. If an organism cannot cope with its physical environment, it will die, so the "analysis" goes—and probably deserves to die. What explanation could be less interesting scientifically and less satisfactory? Very few extinctions recorded in the geologic past yield any evidence at all of the inferiority of the victims *except* their lack of survival. The dinosaurs had been doing very well for something like 140 million years and then, over a fairly short time (the exact amount is debatable), they died out completely. We have caricatured dinosaurs into rather dumb, ungainly creatures, but to label them so is very anthropomorphic. The mammals did not suddenly evolve to push them off the Earth: mammals had been coexisting with dinosaurs for upwards of the 140 million years. Dinosaurs had earned their right to survive. We can invent many a scenario for the demise of dinosaurs but we really know only that they didn't survive.

Another serious problem in thinking about extinction stems from overuse of the Darwinian paradigm. Charles Darwin's major contribution to biology was to propose a mechanism of adaptation based on what he called natural selection, later termed "the survival of the fittest." This process obviously works. Farmers knew long before Darwin that selective breeding could be used to "evolve" new strains of plants and animals. The farmer can eliminate, or prevent from breeding, the undesirable individuals and thereby cause a change in the species that is heritable. This is the "phyletic transformation" I mentioned earlier. It does not result in true extinction and does not help explain the elimination of all breeding populations of a species—or all the species of some larger biologic group.

One way around the problem is to apply Darwin's natural selection at a higher level: *between* rather than *within* species. If species A has more food than species B, or runs faster or withstands cold or heat better, or whatever, then given enough time species B will lose out and go extinct. This is an appealing idea and fits well with the traditional Calvinist views that many of us grow up with. It describes a fair game where goodness triumphs in the end.

The great satirist, Will Cuppy, put it this way:

> The Age of Reptiles ended because it had gone on long enough and it was all a mistake in the first place.

The extinction process I have just described seems so evidently plausible and compelling that experimental investigation and proof seem hardly necessary. And this, I think, is the most likely reason why so little scientific energy has been spent wondering about the hows and whys of extinction. But the history of science has shown over and over again that this can be a most dangerous course. Ideas that seem too obvious to look into seriously often turn out to be dead wrong!

Fortunately, there are some beginnings of serious study of species extinction, both as a biological process in the here and now and as it impacts on the larger scale of the evolution of life. Consciousness among evolutionary biologists and paleontologists has been raised to the point where some of the problematic aspects at least have names. The differential extinction of species is now called "species selection," and paleontologists are beginning to differentiate "background" and "mass" extinction.

For example, my colleague David Jablonski has recently shown that the biological selectivity at mass extinctions is significantly different from that during the background times between mass extinctions. Working with large samples of mollusc fossils of Cretaceous age, Jablonski demonstrated that during the "quiet" part of the Cretaceous, species with floating or swimming larvae had lower extinction rates than the species that brooded their eggs. But during the terminal Cretaceous event, this selectivity disappeared and both kinds of molluscs suffered equally.

In addition to a new awareness of extinction among paleontologists, people around the world are becoming concerned about endangered species and the problem of contemporary extinction, especially in tropical rain forests. It has been in ecology that our general ignorance of the extinction process has been dramatized. We know that if all the habitats of a species are eliminated, the species must go too, but note that this is just another of the old tautologies. There is beginning to be serious research on this and other aspects of the total problem.

DEATH OF THE DINOSAURS

In spite of the drama normally associated with dinosaurs and their extinction, it was of less importance than our culture has assigned it. Many dinosaurs were very large, and some were certainly ferocious, but they never "ruled the Earth" except in the very restricted sense that some of them were top carnivores in certain terrestrial communities. They did not rule the Earth any more than lions today rule the Earth.

At their acme, dinosaurs had perhaps as few as fifty species living at any one time. This is a trivial number when compared with the millions of species in the total global biota. Even among land vertebrates, the numbers are not especially impressive. There are about 5,000 species of mammals living today and about an equal number of reptiles. These proportions were somewhat different in the Mesozoic when the dinosaurs lived, with more reptiles and fewer mammals, but no matter how the bookkeeping is done, those reptiles we call dinosaurs were a minor part of the biology of the Mesozoic era.

It has even been suggested that dinosaurs went extinct more through bad luck than bad genes. With independent species extinctions going on all the time, each species has a certain probability of going extinct in a given time interval. Perhaps with so few species, the dinosaurs as a group died out just because of the unlucky break of all species dying out for different reasons at about the same time. This would be analogous to all players on a baseball team going hitless for an entire game because of individual bad luck rather than some fundamental weakness of the team. This question has been tackled mathematically for the dinosaur case, and it appears that they were not just the victims of bad luck in the sense described. The demise of the dinosaurs in a short time is statistically off scale.

But this does not mean that the dinosaurs as a group did something wrong in the sense of being poorly adapted to their environment, unless one follows the tautological approach of insisting that just because they died they must have been poorly adapted.

The standard and conventional explanation of dinosaur extinction is that their environment (biological or physical) deteriorated gradually during the last five to ten million years of the Cretaceous to a point of no return. Climatic cooling is the most common culprit named as the immediate cause. Indeed, the climate of the Earth was in a cooling phase during the late Cretaceous, and dinosaurs were generally confined to the warmer regions. Also, the number of species of dinosaurs did decline over the later parts of the Cretaceous, so that there may have been as few as twenty-five coexisting near the end.

The climatic-deterioration scenario is certainly plausible and may well be correct. The problem is finding anything resembling proof. A well-known trap in the analysis of historical records is what the famous Oxford statistician, G. Udny Yule, called the "nonsense correlation." Most things that we can measure over time are changing, be they mean temperatures on the Earth or stock-market averages or skirt lengths. Thus, if we look at any pair of such records, they are very likely to be correlated (positively or negatively): either both are going up, both are going down, or one is going up and the other going down. Professor Yule used several decades of British records to show that life expectancy in the general population went up consistently as the membership in the Church of England declined. And he pointed out that it was nonsense to suggest that public health in Britain improved because of declining commitment to organized religion—or the reverse. Using "nonsense" logic, who would argue that the strong correlation between upward trends both in the incidence of lung cancer in America and ownership of pop-up toasters indicates cause and effect?

So, the observed correlation between climatic deterioration and dinosaur decline is suggestive but does not really tell us much. In the context of Nemesis, a more vital question is how long the extinction took. Are we dealing with something that happened over a weekend, as the Nemesis scenario suggests, or was it something that took millions of years, as the climatic interpretation suggests?

One argument offered repeatedly by the Nemesis detractors is that because the ultimate extinction of the dinosaurs was preceded by several million years of decline in number of species (from about fifty to about twenty-five), the stresses that did them in must have been long-term and progressive.

And the Nemesis detractors also argue that the decline in numbers rules out a sudden event like a comet impact, on the grounds that the dinosaurs could not possibly have anticipated the event millions of years before. Although this argument is appealing at first, it does not hold water.

It is hard to find any time in Earth history when some fossil groups are not either waxing or waning. Like skirt length and church membership in our own times, things are generally going up or down. So, to note that dinosaurs were in decline long before their extinction does not in any way rule out sudden catastrophe—or even say much about the nature of whatever environmental stress did them in. In fact, if the Earth was subjected to a sudden and unusual environmental stress at some instant 65 million years ago, we would expect that the biologic groups already reduced in numbers for other reasons would be preferentially killed off.

As I hope is clear by now, we don't know why the dinosaurs went extinct when they did, and we don't know for sure how long the process took. One reason the idea of comet impact is appealing is that such an event leaves enough other signs that it should be possible to pinpoint the event.

The situation of the last days of the dinosaurs is made yet more difficult because the fossil record is rather scrappy. Most dinosaurs lived on land in settings that undergo much more erosion than sedimentation, so that burial and fossilization is a rare occurrence. Because of the fragmentary fossil record, we cannot just "look" at the rocks in the last interval of the Cretaceous period to get the answers.

OTHER CRETACEOUS EXTINCTIONS

The end of the Cretaceous is among the five most severe mass extinctions known in the geologic record. And, as we have seen, the dinosaur extinctions alone were not profound enough to merit mass-extinction status. Actually, organisms were affected in all environments, aquatic and terrestrial. We get the most complete picture from oceanic records, because the marine sedimentary sequence is more complete and fossilization is somewhat more likely on the sea bottom than on land.

The marine fossil record indicates severe extinctions near the end of the Cretaceous, although the severity depends on one's scale. There were about 790 taxonomic families of fossilizable marine animals living at that time. Some of the families probably had only one species while others had scores or thousands of species. The average family probably had about 100 species living at any one time, although this is just a statistical mean of a highly variable array of family sizes.

Of the 790 families of marine animals present in the last conventionally recognized unit of the Cretaceous (the Maestrichtian), 120, or about 15 percent, were extinct by the end of the Cretaceous. The figure for the taxonomic level of genus is approximately 50 percent. The fact that the extinction rate for genera is higher than for families is interesting and actually says something about the nature of the extinctions. If some families were completely immune to whatever happened environmentally and others were totally susceptible, the two percentages should be the same. Some families (with all their genera and species) would be killed completely and the others would be left unscathed.

But this is not what happened. Many families lost species and genera but still survived with at least one species. A good example of this in the terrestrial realm is the fate of mammals in the late Cretaceous. We tend to think of mammals as survivors, but this is true only in the sense that the lineage survived. Many mammalian groups were hard hit and lost most of their species. The marsupial mammals suffered profound losses and nearly died out, but they managed to sneak through with a few species, which in turn later radiated to

produce the very respectable diversity they now enjoy in
South America and Australia.

It is important to try to estimate the late Cretaceous kill
at the species level. This is a difficulty because the fossilization of species is spotty and some mathematical inferences
have to be made. In any event, the best available estimates
are that between 60 percent and 80 percent of marine species
died out. This is not quite as high as the 96 percent estimate
for the Permian mass extinction, but it still approaches complete disaster and suggests conditions I am glad I did not
witness.

It would be nice to give a simple list of the victims and
survivors of the Cretaceous event. This is not so easy to come
by, because the 120 marine families that died out include
some rather obscure animals with esoteric names. We can
make a few generalizations, however. Tropical reefs were
hard hit, especially those that were built at that time by a
rather weird clam called a *rudist* (which looked like an oyster
pretending to be a coral). Also, zooplankton lost heavily,
especially in tropical waters.

On the survivors' side, reef corals themselves got through
the crisis pretty well, as did most deep-sea animals. In all
this, we can see some patterns of selectivity on the basis of
membership in biologic groups, which implies certain
anatomical and physiological common denominators, and on
the basis of broad habitats, such as the ocean surface or the
deep sea. But one of the challenges for future research is to
do a much more thorough job of identifying the winners and
losers, so that we have a better chance of learning exactly
what environmental stresses were responsible for the disaster.

OTHER MASS EXTINCTIONS

A few of the other extinctions were more intense than the K-T event, but most were less. All of them show some selectivity by anatomy or habitat, although the patterns differ from one extinction to the next. It should be possible to say a lot more about this in a few years, because a number of excellent, statistically inclined paleontologists are working on the problem. And data on extinction are now computerized, at least for genera and families of marine animals, so that we are able to ask and answer some of the simpler questions after a few minutes at a computer terminal. The family level history, highly codified, fits on one double-sided, double-density floppy disk, and that for genera takes up about ten disks and is growing. I will have a lot more to say later about what we do with these data, especially on the question of whether major extinctions have a clocklike spacing in time.

One other extinction event of the geologic past has special interest. This happened in the latest Pleistocene epoch between 7,000 and 10,000 years B.C. and was most pronounced among large mammals in North and South America. Mammoths, mastodons, horses, camels, sloths, sabertooths, and a considerable number of other large animals once thrived in North America but died out there in a rather short interval. The kill rate among species was about 70 percent. This is high, but because the killing was almost completely restricted to large, terrestrial mammals, it does not qualify as a mass extinction in the usual sense.

The causes of this extinction have been debated hotly for many years. Climatic change, of course, has been urged by some, and the late Pleistocene was indeed a volatile time for climate. Or was it some purely biological collapse of an integrated eocsystem? By far the most intriguing idea, and the one for which there is mounting evidence, is that the large mammals were eliminated by *Homo sapiens*. The timing is about right for the migration of early man from Asia to North America, and also, hunting sites have been found to contain the remains of large mammals. This may have been the first man-made extinction.

4: Gubbio and the Iridium Anomaly

WHY GUBBIO?

Luis and Walter Alvarez, father and son, working in conjunction with two chemists, Frank Asaro and Helen Michel, found an anomalously high concentration of the rare element iridium at Gubbio in northern Italy. Because the anomaly was precisely at the Cretaceous-Tertiary boundary, and because iridium is normally almost absent from the Earth's crust but relatively common in some types of meteorites, an extraterrestrial interpretation for the Cretaceous extinctions was reasonable.

Was this discovery a matter of good luck or good planning? There are many answers to this but one thing is sure: it is not an example of the stereotyped procedure in scientific research. The Alvarez group did not start out by formulating a hypothesis about the causes of extinction and then going to Gubbio to test the hypothesis.

Walter Alvarez is a Columbia- and Princeton-trained geologist who has worked for a number of years on problems of the deformation of rocks. His approach is a combination of detailed field study and thorough laboratory and theoretical analysis. In the late 1970s, he was working in northern Italy on a thick section of sedimentary rocks, mostly limestones, of late Cretaceous and early Tertiary age. The rock sequence was already well known in the sense that the fossil-based chronology had been worked out. But the chronology by itself does not say anything about amounts of elapsed time (in years). Fossils are invaluable in putting together a sequence of events but little use for saying anything about elapsed time, because rates of evolutionary change are too irregular and too unpredictable.

For his geological project, Walter Alvarez needed estimates of elapsed time. He needed to know whether one particular limestone or clay unit took more or less time to deposit than another. He was also interested in how much time was represented by a thin clay layer at the K-T boundary. This might say something about the duration of the Cretaceous-Tertiary transition.

To get actual dates (in millions of years before present), it would have been theoretically possible to have samples at closely spaced points in the section analyzed for radioactive isotopes. But the accuracy of these methods is generally not good enough to differentiate ages within a short span. Also, Alvarez's rocks were not the best candidates for radiometric dating and the analyses are expensive.

So, Walter and his father came up with an ingenious approach to the problem. A steady rain of meteoritic material falls on the Earth's surface. The big meteorites that make the newspapers contribute some of this extraterrestrial material, but more important in this context is the rain of tiny particles sometimes called micrometeorites or meteoritic dust. The amounts are extremely small but the rate of fall is reasonably constant. Some deep-sea areas, too far from land to collect much continental sediment, do have measurable quantities of meteoritic dust. The farther from land, the greater the concentration, because the dust is less diluted by land-derived material. Therefore, and this is the important part, the relative amount of meteoritic dust in a sediment is a measure of rate of normal sedimentation. This implies a simple equation:

$$\text{percent meteoritic dust} = \frac{\text{(rate of meteoritic dust fall)}}{\text{(rate of normal sedimentation)}}$$

The equation can be rearranged to give:

$$\text{rate of normal sedimentation} = \frac{\text{(rate of meteoritic dust fall)}}{\text{(percent meteoritic dust)}}$$

Therefore, if the rate of meteoritic dust fall is known, normal sedimentation rate (expressed perhaps as centimeters per 1,000 years) can be calculated from the percentage of meteoritic dust found in a sample of the total sediment.

All this seemed feasible to Walter and his father, but the problem was to measure the percentage of meteoritic dust in the Italian rocks. The particles of meteoritic origin are not easily distinguished from normal clay particles eroded from the continents. Here is where Luis Alvarez's contributions became important. He saw the necessity to find some easily measurable property of meteoritic material that could proxy for the actual counts of micrometeorite particles. They decided to analyze chemically for the concentration of the trace element iridium, because of its near-absence in ordinary material of the Earth's crust and its presence in meteorites.

So, with the help of the analytical facilities at Lawrence Berkeley Laboratory and the expertise of Frank Asaro and Helen Michel, a series of the Italian limestones and clays was analyzed for iridium. The result was supposed to give Walter the indicator of rates of sedimentation he needed in the first place.

The surprise, of course, was that the samples analyzed from the small clay layer at the Cretaceous-Tertiary boundary showed far more iridium than was expected or than was found in rocks above and below the boundary.

Many scientists would have dropped the iridium project at this point. The anomalously high concentration of iridium could have been taken to indicate simply that the chemical method was not going to work. The iridium content was too erratic to be useful. Perhaps other aspects of sedimentation or post-depositional history were messing things up. Perhaps the concentrations of iridium in these rocks were naturally too low for the kinds of discrimination Walter Alvarez was seeking. After all, the highest iridium concentration found at Gubbio was only about 10 parts per billion, and there is always a certain amount of laboratory uncertainty. On the other hand, the background levels of iridium above and below the K-T boundary were much lower, about three-tenths of a part per billion, so the 10-parts-per-billion figure really was anomalous.

THE 1980 BOMBSHELL

The lead article in the June 6, 1980, issue of *Science* detailed the iridium story and proposed a large-body impact at the end of the Cretaceous as the cause of the extinctions of dinosaurs and other animals. The paper was written jointly by Luis Alvarez, Walter Alvarez, Frank Asaro, and Helen Michel. This was not actually the first news of the research. A preliminary report had already been presented orally at a meeting of the American Association for the Advancement of Science (AAAS) in San Francisco and the word was getting around. But for most people in the scientific community, this was their first exposure to the idea.

The article itself was unusually long for *Science* (fourteen tightly packed pages in a three-column format) and contained much more than I have covered in my brief summary here. Several other chemical elements were analyzed to test

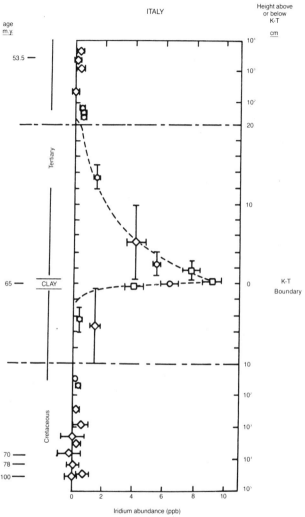

Iridium curve showing the K-T boundary anomaly at Gubbio in Italy. The symbols record the proportion of the element iridium in parts per billion (ppb). One part per billion is about one ten-millionth of one per cent. Time moves forward from the bottom to the top of the diagram, as measured by distance in centimeters below and above the clay layer that marks the transition from Cretaceous to Tertiary limestone. The spike in the curve shows the position of the iridium anomaly. The horizontal error bars indicate the uncertainty in chemical analysis and the vertical error bars show the thickness of rock over which each sample was collected. (After L.W. Alvarez, 1983, *Proc. Natl. Acad. Sci. USA, 80:* 627–642, fig. 4)

for the meteoritic signature. Probably more important, the iridium anomaly was reported from two other localities: one in Denmark and one in New Zealand. This was at least a first step toward demonstrating that the Gubbio results were more than a fluke or the result of some purely local effect.

Several pages in the *Science* article were devoted to the interpretation of the iridium anomaly. Arguments were well reasoned and carefully done, although, inevitably, various aspects of the interpretation have undergone modification since 1980. The basic thesis presented in 1980 was that a large asteroid hit the Earth 65 million years ago and the force of the impact sent up into the atmosphere about sixty times the asteroid's volume in pulverized rock and fragments of the asteroid itself (with its iridium). The atmosphere became so clogged with dust that sunlight was blocked and photosynthesis of green plants was inhibited or stopped completely. This, in turn, broke down food chains and led to the demise of animals dependent, directly or indirectly, on plants for food.

The size of the impacting asteroid was estimated from the amount of iridium found at Gubbio (and elsewhere) and from the known iridium concentrations in meteorites. The diameter of the body was calculated to be ten kilometers (about six miles) plus or minus four kilometers.

INITIAL REACTIONS

Like sediment falling, time has obscured scientific reaction to the 1980 paper. I have some recollection of my own reaction in 1980 and will try to describe it.

I first learned of the Alvarez work when asked by *Science* to read the manuscript as part of its peer-review process. An unusual peer review in several ways, it was being overseen by one of the senior editors, and this meant that the manuscript was especially important or controversial or both. Also, it was clear from my conversations with the editors that there were some problems, stemming at least in part from the unusual length of the piece. I sensed also that the reviews they had already gotten were mixed and that they were looking for additional opinions. All this is fairly common in the peer-review system, and I have never tried to learn the details of this case.

The Alvarez manuscript raised a number of problems for me and I worked pretty hard on my review. On the one hand, the idea of meteorite impacts causing extinctions was familiar to me and very exciting because of the work I had done in Rochester several years before on Öpik's comets and the simulated bombing of Australia. On the other hand, I found technical flaws in the paper. It was also much longer than normal for a *Science* article and I reacted negatively to this. The paper was not particularly well written. I found the style somewhat pretentious. These aspects should not have carried much weight, but I was negative nonetheless.

On the substantive side, I could evaluate only certain aspects of the science—those in areas where I was reasonably competent. This problem of individual limitation has plagued the whole Nemesis Affair. The range of disciplines covered by Nemesis is vast, from paleontology to geochemistry to biology to atmospheric science to astrophysics. No single individual can possibly have more than a superficial knowledge in all the relevant areas. As a result, many mistakes have been made in the debates and discussions of the several hypotheses.

In my review of the manuscript, I found fault with certain aspects and made suggestions for elaboration of the research and for revising of presentation. The details are not important. I ended my review, however, with a fairly unkind statement saying:

SCIENCE

Author **ALVAREZ, L.W. ; et al.**

Title **"Extraterrestrial Cause for the Cretaceous-Tertiary Extinction: Experiment and theory"**

Comments:

The potential impact of this paper has cosmic proportions. If the hypothesis is correct, it will have profound influence on geology and evolution -- not to mention philosophy. The paper will probably change the basic thinking of many people. This is fine if the hypothesis is correct. But if it turns out (later) to be wrong, a lot of damage will have been done and it will take years to recover. Because the hypothesis of the paper goes strongly against conventional wisdom, many readers will do their utmost to find fault.

With these considerations in mind, my comments are directed toward encouraging the authors to fill some gaps in their reasoning and to streamline the paper so that it is the cleanest and least ambiguous presentation possible.

(1) The paper needs more general data on natural variation in iridium content. We are presented with the fact that the boundary clays are much higher in iridium than the overlying and underlying limestones but we are given little or no perspective on iridium distribution in other clays, recent sediments, organisms, sea water, and so on. I realize that data are likely to be spotty but I think the present manuscript understates existing knowledge (see below).

(2) Other clay beds. In the data presented, iridium is high at the Cretaceous-Tertiary boundary but the boundary samples are the only ones from a clay-rich sediment. This invites the reader to ask: If other, non-boundary clays had been analyzed, would they turn out to be just as high as the boundary clays? This is such an obvious question that I am surprised that the authors did not analyze other marly clays. This could be done with museum specimens right in Berkeley as the lithology is very common throughout Mesozoic and Tertiary rocks. If non-boundary clays turned out to be low in iridium, the argument for something special at the Cretaceous-Tertiary boundary would be vastly strengthened!

(Continue on additional sheet if necessary)

Overall Evaluation

(potential)

- [x] Excellent and exciting, merits rapid publication.
- [ ] Above average, publish if space is available.
- [ ] Belongs in a more specialized journal.
- [x] Mediocre or poor, should not be published

(actual) in Science.

If you recommend publication in Science, please check one or more of the following:

- [x] Opens a new and significant area of research.
- [ ] In an established field, rates in the upper: ___1% ___5% ___10% ___20%
- [x] Provides important information of broad interest to the scientific community.
- [x] Is important to specialists in three or more disciplines, namely:

Geology   Paleobiology   Astronomy

Confidential Comments:

This paper will surely be published somewhere. While I do not think SCIENCE should publish it in its present form, it would be feather in your cap to publish it in some form. Therefore, I urge that you do everything possible to convince the authors to revise and rewrite for eventual publication in SCIENCE.

Advisor's Name   Date

Dr. David Raup   1/18/80

The first page of my review of the 1980 Alvarez paper. The form is the standard one used in peer review by *Science,* one of the principal international scientific journals. My review continued with several pages of detailed comments, criticisms, and suggestions. Reviews are normally returned to the author(s) along with the editor's decision on acceptance or rejection of the manuscript. The reviewer's name is generally not revealed to the author(s). For authors, these reviews serve either to explain and support the editor's decision to reject the paper or to help in revision if the paper has been accepted. Note the space for confidential comments (to be seen only by the editor). The Alvarez paper was published by *Science* in June, 1980, with many of the reviewers' criticisms answered, including most of those mentioned on the page shown here.

If a graduate student gave me this manuscript to read, I would see it as a brilliant piece of work (indicating that the student had enormous potential) but I would give it back to be done right!

I thought the work was sloppy and incomplete, however brilliant it may have been. I have never talked with Luis or Walter about my review and presumably my anonymity was protected (until now!). Furthermore, I purposely put the nasty statement at the very end of the review, where the editor could snip it off if he chose—as often happens with vituperative comments in reviews. One doesn't normally compare Nobel laureates to sloppy graduate students.

I have related some of my reactions to the Alvarez paper because I think they say something about the way science works. Deep down, being very uncomfortable with the iridium anomaly and its extraterrestrial interpretation, I struck out at it in some possibly irrational ways. The manuscript did need work when I saw it, but that's what the review process is for. The paper in its published version had many of the holes filled—holes that I, and surely other reviewers, had noted. The paper turned out to be a clear first statement of a hypothesis and the evidence for it.

In the weeks and months following the June, 1980, publication, there was a great deal of discussion of the paper, far more than had ever been generated by the earlier works of Schindewolf, McLaren, and Urey on the same general subject. The comments I heard were almost all negative! Some had to do with the substance of the research, but a surprising number had little to do with details of the evidence and its interpretation. This is partly due to the world I was living in, populated as it was by paleontologists, biologists, and geologists rather than geochemists and astronomers. None of us knew much about trace-element chemistry or meteorites or asteroids. But we did know a lot about the Cretaceous extinctions and the Alvarez study was important to us.

Among the early reactions, a number recurred over and over again. In one form or another they reduced to the following arguments:

1. Not enough is known about the geochemistry of iridium to justify the interpretations made at Gubbio.
2. The iridium enrichment in the K-T boundary clay could be of biological origin. After all, many organisms concentrate certain elements to an extent that, historically, some elements were discovered in marine animals before they were found in sea water itself. The Alvarez paper says nothing about this possibility.
3. Rocks have been analyzed for iridium only close to the K-T boundary. Until we have sampled the entire geologic column, we will not know whether iridium anomalies are unusual. The whole extinction interpretation rests on an assumption of the rarity of the anomaly.
4. How do we know that the iridium anomaly is not just a lag concentrate formed by chemical solution of limestone (containing minor amounts of the element) that was originally above the K-T boundary?
5. If a ten-kilometer meteorite hit the Earth, why hasn't anybody found the crater?
6. The killing scenario based on a dust cloud blotting out the Sun is *ad hoc* and not credible. Why were there no major extinctions among plants at the same time?
7. The late Cretaceous extinctions took place over millions of years, so there is no place for an interpretation based on a single, short-lived event.
8. The Alvarez group contains no paleontological expertise and it is therefore in no position to discuss mass extinction.
9. Mass extinctions are very complex affairs and are the result of many intricate interactions among organisms and between organisms and their environment. A simple expla-

nation for such a complex event as the Cretaceous extinction is inappropriate and likely to be wrong.

10. There is no need or justification to invoke extraterrestrial forces to solve earthly problems. The *deus ex machina* approach was discarded years ago. (Thank you, Mr. Lyell!)

11. The Alvarez group was too quick to call in the press. This is not the way good science is done. Press attention makes the conclusions suspect.

Where did I stand on these arguments at the time? The opinion of one person, not an active participant, is not important, but I can be somewhat more accurate (perhaps) describing my own reactions than trying to divine what other people were thinking several years ago. My recollection of the time is that I was cautiously optimistic about the impact hypothesis. When asked by friends and colleagues, I usually said something like: "I sure hope they are right but the hypothesis has some very serious problems." When queried about the problems, I probably mentioned one or more from the list I have just given—although not any of the last four.

One of the early published reactions of paleontologists to the Alvarez hypothesis was a piece by three prominent practitioners: William A. Clemens of Berkeley, J. David Archibald of Yale (now at San Diego State University), and Leo J. Hickey of the Smithsonian (now director of the Peabody Museum at Yale). Their essay was published in the Summer, 1981, issue of *Paleobiology* under the title "Out with a Whimper Not a Bang." Although it is a calm and thorough piece of scholarship, the authors make no secret of their attitude toward mass extinction at the end of the Cretaceous. Here are a few quotes:

> Paleobiological data cannot rule out the possibility of the occurrence of supernovae, asteroid impacts, or other extraordinary

events. . . . However, analyses of the paleobiological data suggest such an event is not required . . . Moreover, no evidence from any time in earth history conclusively links the collision of extraterrestrial objects with major changes in the patterns of evolution or extinction.

. . . the Cretaceous-Tertiary transition was a period of several tens of thousands if not hundreds of thousands of years in duration, characterized by interaction of a complex of physical and biological factors . . .

. . . the extinctions used to mark the end of the Cretaceous were not the product of one great catastrophe. Biostratigraphic studies . . . reveal that different groups drop out of the fossil record at different stratigraphic levels. . . . This pattern is difficult to explain in terms of a sudden, all-encompassing catastrophe.

About eighteen months after the original Alvarez publication, I wrote a short account of the Snowbird Conference for *Paleobiology* which indicated my thinking at the time. It was very cautious. Although basically favorable toward the hypothesis, I saw what I called "dark clouds" everywhere. Here are some scattered quotes.

There is too little good information on Ir content in the Phanerozoic generally. . . . We do not yet have the perspective that can be provided only by good geochemical data from throughout the record.

. . . diagnosis of synchroneity of iridium spikes and extinctions is problematical.

The effects of an oceanic impact are not clear . . . the whole subject of impact effects is problematical at this point except that there is general agreement that the effects would be large.

> A really serious problem . . . is the fact that the true top of the Cretaceous in seen in only a few sections (mainly in western Europe).

> The whole problem is very messy and I, for one, am not certain where the truth lies.

Another opportunity I have to document my reactions shortly after the Alvarez publication is a paper I wrote in January of 1981, entitled "Extinction: Bad Genes or Bad Luck?" It was prepared for a conference in Barcelona and later published in *Acta Geologica Hispanica.* In this paper, I cited the Alvarez *et al.* (1980) work and described some computer simulations that were asking "What if . . . ?" questions. If mass extinctions are episodic and sudden, what kinds of patterns should we see resulting in the fossil record? I referred to the impact hypothesis as a "claim," which in the jargon of science (and journalism as well) often expresses skepticism. At one point, I wrote:

> . . . Alvarez *et al.* (1980) . . . claim to have hard geochemical evidence for a collision at the end of the Cretaceous between Earth and a 10-kilometer meteorite. Although this event is yet to be firmly documented, it has considerable credibility.

The foregoing gives some flavor of the initial reactions to the Gubbio paper. The rank-and-file response in geology and paleontology was generally negative, but at least the case was being argued. The Alvarez group was getting a hearing unlike Schindewolf, McLaren, and Urey. The troops were angry! But many good scientists were reanalyzing old data and going into the field for new. Several laboratories around the world started analyzing rocks for iridium, and a number of totally different avenues were being followed to find evidence for or against large-body impact at the K-T boundary and elsewhere in the geologic column. In spite of emotions and preconceived ideas, this was scientific research as it should be practiced.

5: *The Three-Meter Gap and Other Evidence*

HELL CREEK, MONTANA

As the impact hypothesis developed over the next three or four years, a few localities in eastern Montana gained international fame. Most of these are exposures of the Hell Creek formation near Fort Peck Reservoir and some have both dinosaur fossils and an iridium anomaly. Of course the question loomed: did the last dinosaur bone coincide with the iridium?

In a way, this is not the all-important question it has been made out to be. As we have seen, dinosaur extinctions were a fairly minor component of the late Cretaceous extinctions. Also, the fossil record of non-marine environments is notoriously fragmented, so Montana with its dinosaur beds was not an ideal place to settle arguments about the causes of mass extinctions. Nevertheless, the dinosaurs did go extinct and their disappearance (and replacement by Tertiary-style mammals) is an important datum in the history of life.

The Montana sites became celebrated in part because their geology and paleontology had been studied by one of the best vertebrate paleotologists in the business: Professor William A. Clemens of the University of California at Berkeley. Bill

Clemens and Walter Alvarez occupy the same building at
Berkeley and the two became friendly antagonists on the
K-T extinction question early in the game.

First analyses favored the Alvarez thesis. The highest and
therefore youngest dinosaur fossil is below the iridium anomaly, as it should be if an impact killed the dinosaurs. But
there are three meters (about ten feet) of sediment in between. To make matters worse, mammal fossils of a definitely
Tertiary aspect have been found between the highest dinosaur bone and the anomaly. To Bill Clemens and many other
paleontologists, the three-meter gap was decisive evidence
that the dinosaurs were dead and gone well before a meteorite hit, even assuming (which they did not) that there was a
meteorite.

Not surprisingly, the Montana situation is more compli-
cated, and the argument between Walter Alvarez and Bill
Clemens, involving many of their colleagues on both sides,
continues to this day (except that Luis Alvarez tells me that
Clemens is willing to reduce the gap to two meters). Space
does not permit much detail but a few of the more important
elements can be mentioned.

I noted earlier that the highest dinosaur fossil is below the
iridium anomaly. This is true only if one specifies that to
qualify as a dinosaur fossil, there must be a close assemblage
of bones from the same animal—a so-called "articulated"
specimen. Odd fragments and isolated bones are ruled out in
this case, because the chances are so good that isolated pieces
may be buried, exhumed by erosion, and deposited again
long after the animal actually died. In eastern Montana,
dinosaur fragments are occasionally found even above the
iridium anomaly, but nearly all geologists agree that these
cannot be used to prove that the dinosaurs were still living.

The iridium anomaly itself is probably a good one, although there has been some argument about it. Things got
a bit confused early on when one anomaly turned out to be

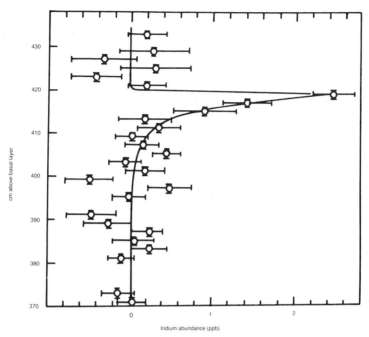

Iridium curve for the K-T boundary in eastern Montana. The scale on the left shows distance in centimeters above an arbitrary datum lower in the Cretaceous. The maximum iridium content is not as high as had been found at Gubbio but still stands well above background. (After L.W. Alvarez, 1983, *Proc. Natl. Acad. Sci. USA*, *80:* 627–642, fig. 10)

due to contamination in the laboratory from a technician's platinum wedding band. Analytical procedures have to be carefully monitored because iridium occurs naturally in small quantities in ordinary platinum, and given the trace amounts of iridium in rocks, accidental contamination can be a hazard. After the wedding-band problem was cleared, the anomaly in Montana was well confirmed.

The field situation around Fort Peck Reservoir leaves much to be desired. The rocks are a somewhat confused sequence of stream and flood-plain deposits. Not all of the critical features occur in a single outcrop. To say the iridium anomaly is younger than the last dinosaur fossils requires a certain amount of inference, although this kind of inference is standard in field geology. And all the geologists who have worked there agree that the anomaly is above the dinosaurs. But there is less agreement on the time of the Tertiary-style mammals that occur above the dinosaurs. Whereas the mammal bones are topographically above the dinosaurs and below the iridium anomaly, geologist Jan Smit has published studies arguing that the mammal remains were deposited in a stream channel that formed after the iridium was deposited, but because the channel cut through the later sediments, the fossils now lie below the iridium. Some geologists agree with Smit's field interpretation and others do not. After a painstaking study of the area, David Fastovsky, a graduate student at the University of Wisconsin, has concluded that it is impossible to decide—although he sees a slight tipping of the evidence against the Smit interpretation.

A stickier aspect of the case is the interpretation of the three-meter gap. Does it really mean that the dinosaurs died long before the meteorite impact? Bill Clemens and a number of other paleontologists have maintained all along that it does. But a counter-argument has been championed, particularly by Luis Alvarez. He notes that dinosaur fossils are not very common even under the best of circumstances, so that a few meters of barren rock is nothing to get excited about. Perhaps the dinosaurs in Montana were alive and well throughout the time represented by the three-meter gap but just did not happen to be preserved. It is not known how much time is actually represented by the three meters of sediment.

One could argue that Luis Alvarez, as a high-energy physicist, has no business making such pronouncements about geology. But people outside a discipline can sometimes be very insightful, in part because they are not steeped in conventional ways of interpreting data in that discipline. To me, Luis made a damned good point.

The paleontologist using fossils to reconstruct geologic history relies heavily on the observed ranges in time of particular species or groups of species. The first occurrence and the last occurrence are the two most important pieces of data. As more field work is done, either in a single area or worldwide, the observed ranges of species tend to become longer just because older or younger representatives are found. For example, the monumental work of the Leakeys and others in East Africa has consistently extended the range of fossil hominids backward in time as more specimens are found. So, the time range observed at any one time is merely the best estimate of the range at that moment.

The range extension effect varies depending on the abundance and kind of fossil being considered. For very abundant and readily preservable organisms, the range is usually not extended much after the first phase of exploration. But for rarely preserved organisms, range extensions can be common and dramatic. The dinosaur situation is somewhere between the two extremes, and Luis Alvarez's argument that we might not be seeing the last dinosaur that ever lived in Montana is cogent. As a physicist, he asked why paleontologists do not routinely give an estimate of the uncertainty for all their time ranges, and he suggested a mathematical approach to how this might be done in Montana. He asked, in effect: What is the probability that the absence of dinosaur bones in the three meters is just a sampling problem? This is a good question that surprisingly few geologists have thought to ask.

The argument over the three-meter gap has been stimulating and valuable but I don't think it has changed many opinions. The mathematical tests proposed by Luis Alvarez have not been implemented fully—and perhaps cannot be in the Montana case. The case remains almost one of asking whether a glass is half full or half empty. To some, the three meters is crucial evidence that destroys the impact-extinction link. To others, the three meters is trivial in comparison to the enormity of the total Mesozoic history of the Earth. In other words, Luis Alvarez the physicist is saying that the impact and extinctions occurred at the same time "to a first approximation" and Bill Clemens the paleontologist is saying that they occurred at very different times. My own inclination is to side with Alvarez, but I have never set foot on the Montana sites and thus am in no position to pontificate.

As I write this in September of 1985, Bill Clemens has just come up with another argument. *Time* magazine reports that he has found a veritable trove of dinosaur remains in Alaska, and he is saying that this disproves the meteorite hypothesis because it shows that dinosaurs could have withstood the long darkness of the arctic winter. A few months of darkness following a meteorite impact would not bother the dinosaurs. We will have to see how this one develops after the evidence is fully digested and published. Go to it, Bill!

The Hell Creek work is not the only research on the impact to be done since the 1980 paper in *Science*. Literally hundreds of geologists, paleontologists, geochemists, and geophysicists all over the world have contributed, and many of the results are impressive.

OSMIUM ISOTOPES

A highlight of the Snowbird Conference of 1981 was a paper presented by Karl Turekian of Yale. Turekian is a very highly respected geochemist and also is known as a keen intellect who is not bound by convention. When Karl talks, people listen. His presentation at Snowbird was not to report what he had done but what he planned to do. Osmium is another of the platinum-group elements that is commonly present in meteorites but extremely rare in ordinary rocks of the Earth's crust. Furthermore, the ratios of the isotopes of osmium differ substantially between the crust and meteorites. Turekian's plan was to analyze the osmium from Gubbio and other K-T boundary sites to see which kind of osmium they contained.

Turekian, an effusive sort of person, was obviously very excited at Snowbird about the prospective osmium work. Also, he made it abundantly clear that he expected to find ordinary crustal isotope ratios and that his study would show that the impact theory was neither necessary nor credible. It was therefore striking indeed when about eighteen months later, Turekian's study was published (in a paper co-authored with J. M. Luck).

The paper reported osmium isotope ratios much closer to those of meteorites than the crust, and Luck and Turekian concluded with a strong statement of support for impact at the K-T boundary. In fact, minor differences in osmium isotope ratios among the several samples led them to suggest that there might have been more than one impact. This aspect is subject to some debate, but I think most geochemists basically support the osmium data as argument in favor of large-body impact.

SHOCKED QUARTZ

A number of years ago, mineralogists learned that the ordinary mineral quartz reacts in strange ways to exceedingly high pressures: odd things, which I do not completely understand, happen to the crystal lattice. This was demonstrated in the laboratory and then found at known meteorite-impact nuclear-bomb sites. It seems that this kind of shock effect can form naturally at the Earth's surface only under pressures of the magnitude experienced with a high-velocity impact of a large asteroid or comet. Two separate minerals called *stishovite* and *coesite,* both forms of quartz, are often associated with the shock metamorphism.

After these relationships were established some years ago, stishovite, coesite, and quartz showing the shocked lattice became important criteria for verifying ancient meteorite craters. And because of this, the number of authenticated craters increased dramatically. My understanding is that the shocked-lattice features are far more definitive than the simple presence of stishovite or coesite, because these minerals are also known to have formed at high pressure deep in the Earth and are occasionally brought to the surface by tectonic forces.

As you can probably anticipate, shocked quartz has been found in the rocks bearing the iridium anomaly. Bruce Bohor and his colleagues at the U.S. Geological Survey in Denver reported finding shocked quartz at K-T boundary sites both in Europe and North America. This was impressive, because the shocked quartz is a tried-and-true indicator of impact and because Bohor's group had no ax to grind and no prior role in the impact controversy.

But the shocked-quartz announcement did not immediately carry the day. One began to hear criticism and complaint almost immediately. I remember a caveat that was passed around much like a rumor. It suggested that the shocked quartz could have originated in the environment where diamonds are formed: *kimberlite* pipes. As a paleontologist with little knowledge of mineralogy, I found this fairly credible, although I did wonder how one got kimberlites that exploded and distributed shocked quartz all over the world. But it was still a point worth investigating. I was soon embarrassed to learn what I should have known: there is virtually no quartz in kimberlites!

The anecdote about diamonds and kimberlites is typical of a lot of the recent debate. Because all of us are operating partially or completely outside the fields of our training, we are sitting ducks for spurious arguments, pro or con. And rumors like the kimberlite one have been flying thick and fast around the scientific community and are having a significant effect on the progress of science. As I write this, another alternative explanation has surfaced, this one based on the occasional presence of shocked quartz in volcanic rocks drawn from the Earth's deep interior. From what Bruce Bohor tells me, however, the quartz in these volcanics shows nothing like the amount of lattice deformation shown by the K-T boundary quartz.

Despite the arguments, the shocked-quartz evidence is potent. I am convinced that were it not for the emotional nature of the mass-extinction issue, the K-T impact question would have been settled completely by Bohor's paper on shocked quartz. Imagine for a moment that extinction was not involved at all. Suppose that the whole research problem were just a question of whether or not a large meteorite hit the Earth 65 million years ago. I submit that under these circumstances, the impact interpretation would have been accepted by all—perhaps on the basis of the original iridium anomalies and certainly with the added weight of the osmium and quartz work. We know that the Earth has been bombarded by large and small meteorites, and the postulated ten-kilometer size is well within reason. We already had a catalogue of about 100 authenticated impact events, many of them much older than 65 million years. So, the 65-million-year event would just have been added to the catalogue and few would have had interest in further research on the case.

In view of this, I am convinced that the negative reactions to the extinction aspect have forced the proponents of the K-T impact to develop a case verging on overkill. Perhaps this is as it should be, because the impact hypothesis of extinction does call for a major shift in thinking about the history and evolution of life. Maybe the proponents of the idea have a special burden of proof.

In any event, the shocked quartz has caused a marked shift in scientific opinion toward accepting a large-body impact at the K-T boundary. But this does not mean acceptance of the link to extinction.

MICROTEKTITES

I mentioned tektites in connection with Harold Urey's research on comets and extinction. Since the time of his work, smaller glassy particles called microtektites have been found in some sedimentary rocks, and these are interpreted, with their larger cousins, to be by-products of meteorite impacts. It was natural, therefore, to look for microtektites at Gubbio and elsewhere near the K-T boundary. Several geologists have in fact found tiny spherules that they interpret as altered microtektites. The spherule composition does not fit, but some aspects of spherule structure do. There remains much argument over whether the spherules were originally microtektites and were subsequently altered to their present composition.

Undeniable microtektites have been found at other levels in the geologic column, even in association with iridium anomalies, but the Cretaceous case remains equivocal. As far as the K-T boundary event is concerned, the microtektite-like spherules are at best subsidiary evidence.

MORE IRIDIUM-ANOMALY SITES

A flurry of chemical analyses followed the initial finding of anomalies in Italy, Denmark, and New Zealand. By the end of 1983, iridium anomalies had been found at more than fifty K-T boundary sites. Reports included the output of seven independent laboratories in the United States, Holland, Switzerland, and the Soviet Union. In some cases, samples from the same sites were analyzed by two or more laboratories.

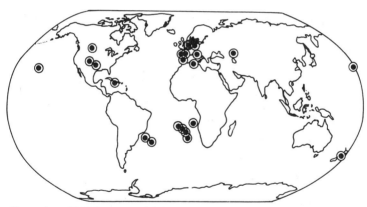

Sites where K-T boundary iridium anomalies were found by mid-1983. They are worldwide in distribution and include the full range of late Cretaceous environments, from marine sediments to swamp deposits on land. It is remarkable that evidence of a short-lived event should have survived erosion at so many different places. The sites shown in ocean areas are based on samples of ancient sediments taken from deep-sea cores. The chemical analyses come from laboratories in Switzerland, Holland, the Soviet Union, and four in the United States (Berkeley, UCLA, Los Alamos, and the Baker Chemical Company). (After W. Alvarez *et al.*, 1984, *Science, 223:* 1183–1186, fig. 1)

Furthermore, the K-T anomaly had been found around the world and in virtually all kinds of sedimentary environments, from the deep sea to swamp deposits on the continents. This made clear that the initial results could not have been a fluke, although many scientists are still concerned about the relatively few analyses that have been done for times other than the immediate vicinity of the K-T boundary.

SOOT AND THE GREAT FIRE

The latest line of evidence to appear on conditions at the end of the Cretaceous is work by one of my chemistry colleagues at Chicago, Edward Anders, and two of his associates, Wendy S. Wolbach (a graduate student) and Roy S. Lewis. They have identified fluffy aggregates of graphitic carbon—in other words, soot—from samples of the K-T boundary clay collected at iridium localities in Denmark, Spain, and New Zealand. The report of this work, with Wendy Wolbach as first author, appeared in *Science* in October of 1985.

The implications of this are very large. The soot was evidently produced by major fires—fires that would dwarf any we have known in historical times. The published report suggests that the soot most likely came from wildfires ignited by the K-T meteorite impact itself. Apparently, even if the impact were oceanic, the heat would be sufficient to start fires more than a thousand kilometers away, The smoke, soot, and other debris could then be distributed worldwide by the winds generated by the fires. The total amount of soot, extrapolated from the amounts found at the three collecting sites, is estimated to be equivalent to more than 10 percent of the world's present biomass! And the effects of the Cretaceous fires are thought to be considerably in excess of those postulated as an aftermath of thermonuclear war by Paul Crutzen, Brian Toon, Carl Sagan, and others working on the nuclear-winter scenarios.

The soot matter is different in a very important way from the other lines of evidence that have been applied to the extinction problem. Iridium, osmium isotopes, shocked quartz, and microtektites were used as *evidence for cosmic impact,* whereas the soot logic *assumes the impact* and uses this assumption to infer the environmental effects. Let me explain. The carbon content of the K-T boundary clay is high but not spectacularly so. The accumulation rate of carbon becomes spectacular *only* if one assumes, as the Anders group does, that the clay layer was deposited in just a few months as fallout from a global cloud. The soot argument thus represents a sort of second-generation theory and is only as good as the impact hypothesis on which it depends.

It is too early to say how the soot report will influence thinking about the K-T extinction, but if the research holds up, we may be much closer to understanding what really happened 65 million years ago.

6: *The Plot Thickens*

THE KILLING SCENARIO

In the original 1980 paper, the Alvarez group suggested that an impact of a ten-kilometer meteorite would clog the atmosphere with dust and other debris and that this would produce darkness at the Earth's surface. They estimated that much of the dust would remain in the stratosphere for several years. Photosynthesis would be inhibited or cut off completely, with resulting disastrous effects on the global biota. Phytoplankton near the ocean surface would die, influencing animals in various ways up the food chain. The 1980 paper made no pretense of providing an exhaustive treatment of either the environmental effects (atmospheric or otherwise) or the biological consequences. After all, the Berkeley group contained no atmospheric scientists, ballistics experts, or biologists. They simply presented a first-cut hypothesis about environmental and biological consequences, based on the data available to them.

Since 1980, a tremendous amount of research has been done by others on the killing scenario. Geophysicists have evaluated the physical effects of the putative K-T impact as best they could. This is not a trivial problem, because of scale. The impacting body was much larger than anything in our human experience—fortunately! Some of the work has been done in the laboratory, studying the cratering and other effects of small projectiles and then mathematically extrapolating to larger size. This is a valid approach but it can be risky: a small error of laboratory observation or in the extrapolation can produce a large error in the result. Some analyses have been purely theoretical, making use of equations developed mostly in engineering for other problems.

One result of this work is a fairly good idea of the immediate physical effects of a large-body impact. The crater would be somewhere in the range of 100 to 150 kilometers in diameter, depending on the impact site. Debris would be propelled aloft at "ballistic" velocities—the debris cloud would blanket the whole Earth essentially in an instant. This is important because it means that the debris cloud would not be restricted to only one hemisphere, as many volcanic and other clouds are. Also, the consensus is that the cloud would be dense enough to reduce light levels enough to cut off photosynthesis.

At this point, atmospheric scientists take over. How long would the dust cloud stay up? Would surface temperatures increase or decrease? These and related questions are still being debated. Not all calculations agree, but the best work I have seen concludes that light levels would remain low for perhaps only about three months. This is less than originally suggested by the Berkeley group but still long enough to have important direct effects on photosynthesis, as well as longer-term climatic effects.

Curiously, there is not full agreement on the temperature effects: some scenarios call for global cooling, possibly initiating glaciation, while others call for warming through a complex chain of events leading to a "greenhouse effect." There are other complicating factors as well. The initial passage of the meteorite through the upper atmosphere might have some strange effects on the chemistry of the atmosphere, including the production of large quantities of nitrogen oxides, which could have environmental effects more devastating than the others being considered.

I should emphasize, as strenuously as I can, that the physical and chemical effects of a large-body impact are completely beyond my training or acquired expertise. I can do little more than read and listen and pass on the flavor of what I have absorbed. The bottom line is that collision with a ten-kilometer body, or even with a one-kilometer body, would be most unpleasant. It may ultimately fall to the paleobiologist to use the pattern of extinctions to find out more about the actual killing mechanism at the end of the Cretaceous, assuming of course that a meteorite really was the culprit.

THE VOLCANIC ALTERNATIVE

Throughout these post-1980 debates, a totally different interpretation has been lurking: *volcanism.* A few first-rate geologists and geochemists have looked at the Berkeley group's evidence and found it wanting. They don't like the basic reasoning of the 1980 paper vis-à-vis preservation of debris from a large-body impact, and they don't believe that the chemistry of the K-T boundary clay indicates meteoritic origin.

Some of the players in the counteraction are Michael Rampino of NASA's Goddard Institute for Space Studies, Charles Officer of Dartmouth, and Charles Drake, also of Dartmouth. Rampino has played a strange role in the extinction story. At the Snowbird Conference in 1981, he gave an iconoclastic but excellent paper suggesting that the high concentrations of iridium were of biological origin. He showed chemical analyses of manganese nodules that had formed on the sea bottom by biologically induced precipitation, and the nodules had high levels of iridium. A couple of years later, however, Rampino was part of a group that found quite high levels of iridium in the gaseous components of eruptions from the Kilauea volcano. Later, Rampino became one of the strongest advocates of meteorite impact. To some, these changes of stance do not look good, but I applaud any scientist who reads the data as he or she sees them at the moment.

The iridium found at Kilauea is seen by some to be a crippling blow to the impact hypothesis. Some volcanos, especially in Hawaii, are known to draw lava from the Earth's mantle—below the crust. The question is whether the Cretaceous iridium is of mantle origin. Papers by Officer and Drake have argued strenuously that the overall chemistry of the Cretaceous clays bears a volcanic signature rather than a meteoritic one. Who is right? Officer and Drake are first-rate scientists and must be listened to, but so are many on the other side. To me, much interpretation of tables of chemical analyses is surprisingly subjective. For perhaps valid reasons that I fail to understand, geochemists have never developed hard-nosed statistical techniques for deciding whether two sets of analyses are the same or different. So, we are left with some people saying that the iridium-bearing rocks of the terminal Cretaceous are clearly like meteorites and other people saying they are definitely volcanic.

The general volcanic interpretation of the K-T event implies a period of devastating volcanism, the like of which we have never seen—that is, if it is to cause the extinction of more than half the animal species on Earth. This seems incredible, but the human species has been around only a very short time and we have no real basis for saying that the level of volcanic activity we have experienced is typical.

Is there other evidence of unusual volcanism about 65 million years ago? In fact, there is. An immense area in India is blanketed by thick basalts called the Deccan Traps. The Deccan volcanism went on for several million years, but nearly all estimates of the starting time fall close to 65 million years B.P. (before present). A few other such basalt flows are known on other continents and of different ages, but because we have never actually witnessed this type of eruption, very little is known about environmental effects. It is not known for sure whether the environmental effects would be local or worldwide, or whether they would involve fundamental properties of climate and atmospheric chemistry.

As a paleontologist, I could shrug off the volcanism-versus-impact problem. After all, what difference does it make what conditions were like for the biota as long as they were bad? But the question is obviously more interesting and serious than that, and my colleagues and I have been listening to the arguments with great interest. My hunch at the moment is that the evidence for meteorite impact at the K-T boundary is much stronger, but the returns are not all in.

A possibly delicious irony of the volcanism-impact debate is that both sides might be right. One longstanding problem with the impact hypothesis is the location of the crater. The critics have been asking about this from the start and the answers have been cast in probabilities. The crater should have been a big one, perhaps 150 kilometers in diameter. If the meteorite had landed in the ocean, its crater would have stood a good chance of having been lost by now, because the sea floor is continually being consumed by what is called *subduction* at the continental margins. If the impact had been on land, the crater would have been subjected to 65 million years of erosion and might well not have survived in recognizable form. It is agreed that the 100 or so craters that we do have are few in comparison to those that have succumbed to subduction and erosion.

Another, perhaps wild, explanation for the lack of a K-T crater is that the meteorite landed in India, touched off the Deccan volcanism, and was subsequently covered up completely by lava flows. This idea has kicked around informally for two or three years, and I do not know whether geophysicists, whose business it really is, take it seriously. They will have to tell us whether a ten-kilometer body colliding with the Earth (perhaps with the high velocity of a comet in retrograde orbit) would have the force to punch through the relatively thick continental crust to initiate a major volcanic episode.

A minor diversion in the history of the volcanic interpretation occurred shortly after the 1980 Alvarez paper. It was suggested in a "Nova" film called "The Death of the Dinosaurs" that the impact point might have been Iceland, thus producing the volcanism that has characterized Iceland since its inception and that produced the island itself. The problem, which was promptly pointed out by many of the geologists who saw the program, is that Iceland is many millions of years younger than the end of the Cretaceous. So much for good ideas.

IMPACTS AT OTHER MASS EXTINCTIONS?

Not surprisingly, quite a number of people launched programs to analyze for iridium in other parts of the geologic column. Everyone realized that the K-T boundary anomaly would have real significance only if iridium anomalies turned out to be rare elsewhere in the record. Also, it was only natural to wonder whether other mass extinctions were associated with an extraterrestrial signature.

Chemical analysis for iridium is time-consuming and costly, and this greatly limits the number of samples that can be processed. As a result, most of the analyses have come from points in the geologic column where mass extinctions are known to occur. This is not the best science, but scientists are human and compromises have to be made.

Only one person that I know of has done a truly systematic series of analyses over a long span of time without regard to the presence or absence of mass extinctions. This is Frank Kyte, a graduate student at UCLA. He has laboriously sampled deep-sea cores that cover a long time span, from just below the K-T boundary to the present. His procedure is rather coarse, so he may have missed some minor events, but he is able to detect an anomaly of the magnitude of that at Gubbio. My understanding is that he has completed the sequence and has found only the expected K-T boundary anomaly and a very small one, presumably local in extent, near Antarctica.

There is, however, a bright cloud on the analytical horizon. A group at Lawrence Berkeley Laboratory, headed by Luis Alvarez, has designed and is building a totally new instrument for measuring iridium content in rock samples. When operational and fully automated, this instrument will be able to process 20,000 samples per year, operating twenty-four hours a day.

Little numbered containers, each about the size of a camera battery, will be filled and sealed in the field, sent to Berkeley, and put into a hopper. The machine will take the containers one by one, read the number, do the analysis, and print out the results. When I saw the prototype under construction at Berkeley a few months ago, one of the stickiest problems was trying to figure out a foolproof method of getting the instrument to read the numbers of the little containers. I presume this problem has been solved and that we will soon have the comprehensive data on iridium in the geologic record that we so badly need.

In the meantime, existing facilities are being used at Berkeley and several other laboratories to extend the data base. To date, iridium anomalies have been reported from five other times of major extinction. All of them have problems, and considerably more work will be needed to verify their significance.

1. *Eocene-Oligocene boundary* The Eocene and Oligocene

are two of the main time units of the Tertiary period, with the boundary between them being about 38 million years B.P. A real but rather small anomaly has been found there at a number of sites around the world. It was not picked up by Frank Kyte's low-resolution analysis. A particularly interesting aspect of this case is that the iridium is associated with abundant, unaltered microtektites. If the microtektites are taken as sure evidence of large-body impact, as they are by almost everyone, then we have a nice conformity of two independent lines of evidence.

The intensity of the extinction at the Eocene-Oligocene boundary is somewhat in the eye of the beholder. Some paleontologists see it as a major event, while others find the extinction fairly trivial. The best record of the extinction is in cores taken in deep-sea sediments. In these, the association between the iridium, the microtektites, and the disappearance of quite a number of microfossil species is striking. But the disappearance of a few microfossils does not make a major extinction, because, as we have seen, species enter and leave the geologic record all the time.

At a larger scale, twenty-eight marine-animal families (about 3 percent) go extinct somewhere in what is known as the Upper Eocene, an interval of about four million years preceding the Eocene-Oligocene boundary. Whereas this is nothing like the extinction rate in the last unit of the Cretaceous (15 percent of families), it still stands significantly above background levels. The critical question is where the twenty-eight family extinctions are within the four million years of the Upper Eocene. Much more detailed work will be needed before this question can be answered satisfactorily.

Even more curious about this case is that there are apparently two or more microtektite layers, although only one has an iridium anomaly. This has led some observers to say that the microtektites may not be a good indicator of meteorite impact, but to others the multiple microtektite layers suggest a shower of meteorites over a period of time. As we will see when we get into the heart of the Nemesis Affair, the possibility of showers of meteorites is a serious proposition.

2. *Jurassic* This is a very curious one. In the Jurassic period, the time unit preceding the Cretaceous and spanning the time from about 210 to 145 million years B.P., there were a number of smaller extinction events, several of which have been used to mark subdivisions of the Jurassic. One of these is at the boundary between the Callovian and Oxfordian stages, which also happens to separate what are called the Middle and Upper Jurassic.

At this boundary, some very curious things seem to have happened. In most localities in eastern and western Europe, there is quite a bit of missing section, with evidence of erosion and chemical solution of the older, Callovian rocks. Much of the best work on these exposures has been done by a Polish geologist, Wojciech Brochwicz-Lewinski of the Geological Institute at the University of Warsaw. Working with a number of Polish, Spanish, and other colleagues, Brochwicz-Lewinski has scoured the outcrops to try to find evidence of what happened at the end of Callovian time. And he has found some interesting stuff. Among other things, he unearthed small magnetic spherules that are clearly of cosmic origin. Similar spherules have been found at Tunguska, the site of the 1908 comet event in Siberia.

Naturally, Brochwicz-Lewinski wanted to look for iridium, but he does not have the laboratory capability in Warsaw. I remember an amusing incident in this connection. In August of 1984, Brochwicz-Lewinski sent me three samples from the Polish Jurassic. We had met at a dinner party in Warsaw some years before, and he hoped that I could forward the samples on to Frank Asaro at Berkeley for analysis. I did this, but they were not analyzed immediately. Frank had many samples from different ages and places and the Polish material had to wait in line.

Then, Brochwicz-Lewinski visited Jack Sepkoski and me in Chicago for a couple of days in May of 1985, and we were anxious to know what Frank was finding in Berkeley. I called Frank. He said he could run one of the samples overnight and maybe have some preliminary results by the end of the next day. But which of the three samples was most likely to have iridium? Wojciech chose one and the analysis started. We had a lot of fun the next day waiting for the results. This was the sort of stereotyped, rah-rah kind of science that few of us ever actually see.

At almost the last minute before Wojciech had to leave Chicago, we got the critical call from Berkeley. Just before, Wojciech had written on a scrap of paper his guess of what the iridium concentration would be. I liked the bravado of my Polish colleague but did not attach much importance to the number he had written. When we got the results, there was indeed an iridium anomaly, and it matched the number on the scrap of paper within experimental error. Only then did Wojciech explain. Splits of the same specimen had already been analyzed by Intercosmos, the Soviet space agency, and Wojciech knew the answer all along. But his action was completely sensible. He wanted independent corroboration, and he did not want to influence any of us with preconceptions. A good trick. I learned from this experience that the Polish scientists have an advantage over the rest of us because they can work as easily with the Soviets as with the West.

The true significance of the Jurassic anomaly and its associated cosmic spherules is not clear yet. Because there is always a slow but steady rain of cosmic material on the Earth, including meteoritic dust with trace quantities of iridium, the Jurassic case could be explained as a long-term accumulation of extraterrestrial material in the absence of other sedimentation. Remember that the critical boundary in Europe is characterized by a substantial time gap. So, a meteorite impact may not be needed to explain the extraterrestrial material.

Brochwicz-Lewinski is confident that the time gap will not explain the anomalies, and I have a lot of confidence in him as an experienced geologist. Furthermore, he thinks the erosion and solution of rocks, creating the gap in the record, may be due to the extreme oceanic conditions produced by a large-body impact. Much remains to be learned. In the meantime, it is interesting to note that the Jurassic iridium is found concentrated in fossil bacteria.

3. *Permian-Triassic boundary* The mass extinction at the end of the Permian, about 250 million years B.P., was the biggest of them all and naturally has been the focus of much attention. The problem with the Permian is that there happen to be exceedingly few good, continuous exposures of latest Permian rocks and of the boundary between the Permian and overlying Triassic. And the best sections are in some very inaccessible places: Iran, northeast Greenland, northern Pakistan near the Afghan border, and in China. You will remember from Chapter 2 that Otto Schindewolf worked on the Permian of Pakistan. The problem with that section is that there is a substantial hiatus in preserved sediment, with as much as several million years of time unrecorded.

The most complete Permian-Triassic boundary sections, by far, are in China. These were inaccessible, even to the Chinese geologists, through the difficult years of the Cultural Revolution, but they are now being worked on, and the Chinese are most anxious to join in the fun and excitement of the search for iridium.

In fact, a group of Chinese physicists, geologists, and paleontologists have reported an iridium anomaly at just the expected spot in the sequence! The group, headed by Dr. Sun Yi-yin of the Chinese Academy, described the anomaly at an International Geological Congress in Moscow in the summer of 1984 and published it simultaneously (in English). So far, several attempts to reproduce these results in other laboratories have failed. Most of us are cautiously but eagerly awaiting further collecting and more iridium analyses.

Why are good Permian-Triassic boundary sections so hard to find? This may be just bad luck in an historical record that has many accidental gaps. But there may be more to it. During the last several million years of the Permian, sea level was going down irregularly but persistently. Large ocean areas became isolated and left salt deposits as the seas retreated.

One result of the lowered sea level is that the presently exposed continents have very few late Permian and early Triassic marine rocks. Many people have even suggested that the sea-level lowering itself was responsible for the Permian extinctions, and they point to other times of extinction that seem to be associated with sea-level lowering. This is, in fact, one of the prime alternatives to extraterrestrial explanations of mass extinction and makes it doubly important to have more iridium analyses from the P-T boundary in China.

4. *Late Devonian* This, you will remember is Digby McLaren's favorite mass extinction and the one he suggested —with no evidence—in 1970 as an impact phenomenon. In October of 1984, a group including Carl Orth (Los Alamos National Laboratory) and McLaren reported in *Science* an iridium anomaly at or near the boundary between the Frasnian and Famennian stages of the late Devonian, about 365 million years B.P. And this, of course, is the mass extinction McLaren was talking about. The samples had come from the Canning Basin in northwestern Australia.

The analytical instruments at Los Alamos are so precise that Orth works in parts per trillion instead of the parts per billion of most other laboratories. The Orth group had been working for two or three years on Devonian samples from

Europe and North American without finding any unusual concentrations of iridium. But the Australian analyses showed about 300 parts per trillion, about twenty times the background level in Australia for rocks above and below.

But here too there have been questions and problems. Why had the anomaly not been found in Europe and North America? More important, the Australian iridium occurs only in a certain kind of fossil bacterium, *Frutexites*. This raises the ugly possibility that the bacteria were concentrating the small quantities of iridium present in their normal environment. We have already seen that biological concentrating mechanisms are a possible alternative to meteorite impact.

The fact that the iridium in Australia is found only in fossil bacteria could also be interpreted to mean that the bacteria contain a lot of iridium just because there was a lot around, analogous to the high levels of DDT in birds and other animals exposed to excess DDT. Did the bacteria concentrate iridium "on purpose" or just because it was there?

The Devonian case, clouded as it is by the biological questions, is still being debated. For some people, it is the final proof of the impact-extinction link. For others, it can be explained easily by non-extraordinary means. I don't know which side is correct. The outcome of the Devonian case will be important for the Jurassic case, as well, because the Jurassic of Brochwicz-Lewinski has the same kind of bacterial preservation of iridium.

5. *Precambrian-Cambrian boundary* This is the weakest of all the iridium anomalies, but it could prove to be the most important. As we go back in geologic time, the lush fossil record of complex organisms stops abruptly at about 570 million years B.P., the base of Cambrian. The transition between the rather simple fossils of the late Precambrian and the advanced life of the Cambrian has long been a puzzle. There may or may not have been a sudden environmental change,

and there may or may not have been an extinction. But it was surely a critical point in the history of life on Earth.

A brilliant geochemist in Zurich, Ken Hsü, has put together considerable evidence for changes in the composition of ocean waters at times of mass extinction. He finds that a number of important relationships, involving mainly oxygen and carbon, are totally upset at these times of crisis. He calls these "Strangelove" perturbations, resulting in a Strangelove Ocean lasting for some thousands of years. And he finds meteorite impact to be a highly plausible trigger for these perturbations in ocean chemistry.

Furthermore, Hsü thinks he has evidence for the Strangelove conditions near the Precambrian-Cambrian boundary in South China. A Chinese group consisting of Messrs. Fang, Yang, and Huang published a paper in 1984 reporting an iridium anomaly at the boundary. But time correlations are very dicey this far back in geologic history and it remains to be seen how the anomaly and Ken Hsü's Strangelove Ocean will hold up.

As I indicated earlier, the five cases of possible impact-extinction pairs other than the K-T boundary all have problems. But they are intriguing problems that will probably lead us closer to final answers about the causes of mass extinction. And this will be true whether or not any of the five turn out to be proven examples of large-body impact.

TWO OPINION POLLS

Scientific questions ought not to be settled by popular vote, but the opinions of groups have an enormous influence on the course of scientific research. In the summer of 1984, two paleontologists made a quasi-scientific survey of about 500 geologists, paleontologists, and geophysicists in Europe and

North America. The pollsters were Matthew Nitecki of Field Museum in Chicago and Antoni Hoffman of Columbia's Lamont Geological Observatory.

The results were intriguing and a bit amusing.

21 percent were convinced that the Cretaceous-Tertiary mass extinction was caused by a meteorite impact (50 percent in a subsample of American geophysicists);

40 percent believed there was a K-T impact but that it did not cause the extinctions;

27 percent did not think there was an impact at the K-T boundary; and

12 percent did not think there was a mass extinction or an impact at the end of the Cretaceous!

The first two numbers add up to an impressive 61 percent who accept the basic evidence for large-body impact at the end of the Cretaceous. This is high when you consider the natural resistance to a non-Lyellian interpretation and the newness of the idea. On the other hand, the 61 percent is surprisingly low when you think of the vast array of independent evidence that had been amassed for the impact by the summer of 1984.

In October of 1985, the Society of Vertebrate Paleontology held its annual meeting in Rapid City, South Dakota, and 118 out of the 300 attending participated in a survey of opinions on the causes of dinosaur extinctions. This was a significant poll because it came after all the additional evidence of meteorite impact I have described here, and because it involved vertebrate paleontologists central to the extinction debate.

The results of this second poll were reported exhaustively in a *New York Times* article on October 29 by Malcolm W. Browne. The questions were phrased somewhat differently from the 1984 poll but the results can still be compared. On the question of a terminal Cretaceous impact:

> 90 percent accepted the evidence for impact, and
> 10 percent denied the impact.

On the question of extinction:

> 4 percent accepted the impact as the major cause of
> the dinosaur extinctions;
> 43 percent accepted the impact but did not think it
> caused the extinction of the dinosaurs; and
> 27 percent felt that there was no mass extinction of
> land animals to be explained.

In comparing the two polls, I think it noteworthy that among the vertebrate paleontologists, probably the most conservative group in the whole impact-extinction business, 90 percent have accepted the evidence for meteorite impact. This is an impressive increase over the 1984 figure of 61 percent and must reflect the mounting geophysical and geochemical evidence. On the other hand, the number of vertebrate paleontologists supporting a link between impact and extinction is vanishingly small at 4 percent. The main argument heard at the Rapid City meeting was one we have heard before: the dinosaurs had been in decline for a long time before the meteorite struck.

Also, by the fall of 1985, one sensed increasing anger on the part of opponents of the meteorite scenario. For example, the *Times* article on the Rapid City meeting includes the following quote from Robert T. Bakker, a dinosaur expert at the University of Colorado Museum and one of the principal originators of the warm-blooded dinosaur theory:

> The arrogance of those people is simply unbelievable. They know next to nothing about how real animals evolve, live and become extinct. But despite their ignorance, the geochemists feel that all you have to do is crank up some fancy machine and you've revolutionized science. The real reasons for the dinosaur extinctions have to do with temperature and sea level changes,

the spread of diseases by migration and other complex events. But the catastrophe people don't seem to think such things matter. In effect, they're saying this: "We high-tech people have all the answers, and you paleontologists are just primitive rock hounds."

I find this statement more than a little appalling and I am glad to say that the charge of arrogance is untrue. The last few years of extinction research have brought me into contact with many of Bakker's "high-tech people". I have found them uniformly humble about their ignorance of paleontology and genuinely eager to learn from any "primitive rock hound" willing to give them some time.

7: *Enter Periodic*
Extinction

THE Nemesis theory of the companion star would not exist without the proposition that extinction events occur with clocklike periodicity, every 26 million years. It is thus important to go into some detail on the origins of the periodic-extinction idea and the evidence for it.

FISCHER'S CYCLES

In 1977, Alfred G. Fischer and Michael A. Arthur published a paper with the title "Secular Variations in the Pelagic Realm". This turned out to be a prophetic piece, although we did not know it at the time. Quite the opposite. Many of us (most, in fact) did our best to look the other way. Fischer and Arthur were claiming that the major extinctions of the past 250 million years were evenly spaced, coming every 32 million years. This was anathema!

We all knew that the history of the Earth was too complex to be amenable to such a simplistic description. What would keep the system on time? Biological communities operating on scales of a few months or years could not keep time, so how could the much more complex global system stay on a schedule for hundreds of millions of years? Although simple models for the dynamics of local communities and even whole biotas had been kicking around for years—some of them quite appealing in special situations—they generally did not work when tested against hard historical data. The Fischer and Arthur suggestion was going much too far.

Here is what the paper said and did. Fischer and Arthur had put together the geologic histories of a somewhat miscellaneous collection of attributes, some from the fossil record and others from the rocks. Included were the numbers of species (or larger groups) of various kinds of marine organisms, some information on community structures (such as the presence or absence of large predators at various times), data on sea-water temperature, carbon isotope ratios, and several indicators of sea level. This was not a dispassionate job of data collection and statistical analysis. They chose the kinds of data they thought would be most sensitive to the environmental changes they were interested in, and they made no special attempt to justify their choices.

They then plotted these as graphs of change through time for the 250-million-year span from the end of the Permian to the present day and proceeded to interpret the wiggles and woggles. They thought they saw cycles, with the same sequence of changes repeated every 32 million years. I have seen better cycles in the Dow Jones averages.

To me, the Fischer and Arthur paper was not a step forward. We had been striving to pull paleontological research into the twentieth century by promoting testable hypotheses and good, rigorous science. A testable hypothesis did not have to be mathematical, although that was preferable. Above all, we wanted to get rid of the Kiplingesque practice of writing Just-So Stories on the basis of miscellaneous information and sheer intuition. And when conclusions were drawn on the basis of quantitative data, we wanted no-nonsense tests of statistical significance. In the Fischer and Arthur graphs, the only indication I saw of cycles or periodicity could be explained by the subjective way in which the data were chosen.

I have liked and admired Al Fischer for many years and I would not jump on him like this were it not for the fact that I now think his 1977 paper is a brilliant piece of science even though it is not the kind I like best to do.

In 1977, Al Fischer was a professor of paleontology at Princeton and Mike Arthur was his graduate student. Al has since moved to the University of Southern California and Mike has completed his studies and gone on to the University of Rhode Island. Al Fischer has done many things in his career and has done them well. He has been an exploration geologist looking for oil throughout much of South America. As a young assistant professor at the University of Kansas, he co-authored what became the standard textbook on invertebrate palentology for decades. He has done backbreaking and brilliant field geology in the Alps of his native Austria and he is a world authority on the taxonomy of an obscure group of fossil echinoderms. His main characteristic, however, is that one never knows quite what he is going to do or come up with next. And his work on cycles is typical.

Fischer and Arthur did not claim extraterrestrial forces to drive the 32-million-year cycles, although they considered the possibility of some unknown fluctuation in the Sun's brightness. Rather, they thought the ultimate driving force was within the Earth, having to do with unknown cycles of convection in the Earth's interior.

That Fischer and Arthur were a little vague on mechanisms is understandable. There is no theorem in science that says that the description of a phenomenon (the cycles, in this case) must be accompanied by a mechanism for how the phenomenon works, although a viable mechanism is always preferred. The absence of a mechanism is often used as a weapon against research conclusions that we don't like. For example, the idea of continental drift was put down for many years partly on the grounds that nobody had come up with a mechanism. Later, in the 1960s the fact of drifting continents was accepted because new data on the Earth's magnetic field made it compelling—even though there still was no acceptable mechanism.

THE NASA WORKSHOPS

As I mentioned near the end of Chapter 1, NASA convened a series of workshops starting in July, 1981. The purpose was to explore possible NASA sponsorship of research in the Evolution of Complex and Higher Organisms (ECHO). The Life Sciences Division at NASA has long supported research on the origin and earliest evolution of life, as well as research on the evolution of intelligence. These and other programs are related to the general quest for a better understanding of life in the universe, with an eye toward the Search for Extraterrestrial Intelligence (SETI).

NASA is an uninhibited agency willing to try new things and the ECHO workshops were no exception. At the workshops, an exciting bunch of people from disciplines unused to communication gathered to brainstorm the history of advanced life. In the group, there were one or more geneticists, botanists, zoologists, geochemists, geophysicists, astrophysicists, atmospheric scientists, oceanographers, and of course paleontologists and geologists. A philosopher or two would have sweetened the pot. Several of the people who were to become important in the Nemesis Affair were in the group, including Al Fischer, Bill Clemens, and Jack Sepkoski.

The first workshop occurred while the Alvarez hypothesis of mass extinction was still being absorbed. Meteorites were very much on all our minds. But we wanted to do a broad job, and although we spoke of meteorite impacts and their possible effects on global biology, we also worked with as many other aspects of the Solar System and Galaxy as might be relevant. Supernovae, changes in the Sun's luminosity through time, changes in the Earth-Moon system, and possible effects of our passage through galactic density waves: all were potentially important to life on Earth.

We worked with periodic phenomena driven by extraterrestrial forces, but at time scales of thousands rather than millions of years. It had recently been established that Milankovich cycles, regular changes in the orbits of the Earth-Moon-Sun system, can and do affect climate on Earth in a complex of cycles of 22,000, 41,000, and 100,000 years. This was exciting!

One research recommendation to flow from the workshops was a plea to study longer cycles, but this was not stressed. Al Fischer talked about his 32-million-year periodicity as well as some newer work he was doing on shorter, Milankovich-style cycles spotted in older rocks. All of us knew about the 32-million-year business, but I think we were embarrassed by it. Al Fischer was a fine scholar but the big cycles did not look good.

As the workshop series drew to a close in the spring of 1982, it was time to put together a report to advise NASA headquarters, and anyone else who would listen, on research policy. The report, since published, contained many aspects and pregnant ideas, but in the context of our subject, most interesting is the absence of Al Fisher's 32-million-year cycles. All of us worked with editing and revising the report manuscript for many months. Pieces were added and removed through countless drafts. Although I no longer have the early versions, I distinctly remember that Al's cycles were there along with a reproduction of his original plot of wiggly lines. But the 32-million-year cycles were completely expunged by the editing process. Much of Fischer's other work is discussed, of course, but not those big cycles. An interesting example of peer review in action. As chairman of the group, I will confess to a large role.

SEPKOSKI'S COMPENDIUM

J. John Sepkoski, Jr., is a paleontologist at Chicago who was trained by Stephen Jay Gould at Harvard. Jack is both brilliant and hardworking, and it is the latter attribute on which Nemesis is ultimately based. For years, Sepkoski has had a sort of hobby of compiling data on the fossil record of life. His field area is the library and its thousands of reports and monographs that record fossils found since the middle of the eighteenth century. Jack's field area has been a frustration to the television crews and journalists who have been chasing the Nemesis story and have wanted pictures of Jack on a mountainside cracking rocks. He just doesn't do this—at least not for his compilation project. In another mode, he works out of doors on Cambrian rocks, but this sort of hands-on study would be too slow (and redundant) for a global compilation of data.

From the source literature of paleontology, Jack has extracted two kinds of information—first, the most accurate or likely taxonomy of the fossils, and second, the range in time over which each group of organisms existed on Earth. The time range involves just two pieces of data: the age of the first occurrence and the age of the last occurrence. This sort of compilation has been made before, often by teams of specialists, and Jack used the early works as a starting point. He has been able to go much farther in the process, and the result is a uniquely comprehensive, more "noise"-free, catalogue.

One phase of the project was published in 1982 under the title *A Compendium of Fossil Marine Families.* It is a moderately thick and utterly boring volume of names of families, their first and last known occurrences, and literature citations for each. Included are about 3,500 separate taxa: some extinct, some still alive, some ancient, some geologically young. Jack makes no claim that the *Compendium* is perfectly accurate or complete. It is only a stage in developing

Or. Palaeoisopoda
  Palaeoisopididae   D (Sieg)   (23,132,208)

Or. Pantopoda
  *Palaeotheca   D (Sieg)   (23)

Trilobitomorpha

Cl. Trilobita

[Classification primarily from the *Treatise*, Pt. 0, except for the Agnostida which is from Öpik (1963). Stratigraphic ranges are from numerous sources, as indicated.]

?Or. Agnostida (=Miomera)

Agnostidae	€ (Boto)	— Θ (Ashg)	(132,208)
Clavagnostidae	€ (uMid)	— € (Dres)	(132,148,193)
Condylopygidae	€ (Boto)	— € (uMid)	(110,132,208)
Diplagnostidae	€ (mMid)	— Θ (Trem)	(132,148,193)
Discagnostidae	€ (Dres)		(226)
Eodiscidae	€ (Atda)	— € (uMid)	(110,132,193)
Pagetiidae	€ (Atda)	— € (mMid)	(132,148,287)
Phalacromidae	€ (uMid)	— € (Dres)	(171,172)
Sphaeragnostidae	Θ (Ashg)		(226)
Trinodidae	€ (Dres)	— Θ (Ashg)	(226)

Or. Redlichiida

Abadiellidae	€ (Atda)	— € (lMid)	(132,141,395)
Bathynotidae	€ (Boto)	— € (lMid)	(148,193,208)
Chengkouiidae	€ (Boto)		(393)
Daguinaspididae	€ (Atda)		(132,141)
Despujolsiidae	€ (Atda)		(141)
Dolerolenidae	€ (Atda)	— € (Boto)	(141,195,261)
?Ellipsocephalidae	€ (Atda)	— € (mMid)	(110,141,261)
Emuellidae	€ (lMid)		(173,245)
Gigantopygidae	€ (Boto)		(141,173)
Hicksiidae	€ (Boto)		(141)
Kueichowiidae	€ (Boto)		(173,393)
Longduiidae	€ (Boto)		(393)
Mayiellidae	€ (Boto)		(42,173)
Neoredlichiidae	€ (Atda)	— € (Boto)	(141,173,262)
Olenellidae	€ (Atda)	— € (mMid)	(110,132,262)
Paradoxididae	€ (Atda)	— € (uMid)	(110,173,262)
Protolenidae	€ (lTom)	— € (mMid)	(92,141,173,262)
Redlichiidae	€ (Atda)	— € (mMid)	(132,141,173,262)
Saukiandidae	€ (Boto)		(141,262)
Yinitidae	€ (Atda)	— € (Boto)	(195,393)
Yunnanocephalidae	€ (Atda)		(170,195)

*May be related to unlisted extant families.

A page from the Sepkoski *Compendium*. The whole *Compendium*, with about 3,500 fossil ranges, is the basis for the computer analysis of the fossil record that led to the Nemesis theory. Each of the words ending in "ae" is the latinized name of a family of marine animals. The second and third columns give coded geologic ages of the oldest and youngest occurrences, respectively, of species of the family. If there is no entry in the extinction column, the family originated and went extinct in the same time interval. The numbers in parentheses on the right indicate bibliographic sources. Since the *Compendium*'s original publication in 1982, it has been updated and corrected several times, so that many of the original entries, including several on this page, have been augmented or made more precise. (© Milwaukee Public Museum, reprinted with permission.)

a data base. In fact, since 1982, many new references have been found, users have noted errors, and its size has grown by about 10 percent. Jack sends out annual correction lists and updates, so the *Compendium* has become a growing research base for many people.

The *Compendium* was put on computer, latinized names and all, early in 1983 and it now moves around the country on floppy disks. Since 1984, Jack has also been working on the counterpart at the genus level. He is up to about 30,000 entries and going strong.

NUMBER CRUNCHING

What can a paleontologist do with a compilation like the Sepkoski *Compendium?* For many practitioners, not much. These are the research paleontologists who concentrate on a single group of fossil organisms, one slice of time, or one geographic area. For these tasks, the *Compendium* is much too coarse and generalized: the specialist has much better information at hand. But for the generalist, the *Compendium* is a unique resource. How does the evolution of diversity of fish compare with that of the swimming reptiles? How does the rate of family origination for all marine animals change with time? And so on.

The science moves forward by having a good mix of specialists and generalists. Neither can function well without the other, although one might not think so sometimes when the two kinds of paleontologist get to arguing. Fortunately for paleontology today, the mix is good. We have Leakeys and we have Goulds.

For me as a generalist, the *Compendium* was like a new toy. And a much better behaved toy than the primitive data bases I had been working with. In a matter of seconds, with the computer on my desk and a pretty good text editor, I can get the names and affiliations of all marine families that went extinct in any time interval, for example.

In the winter and spring of 1983, Jack and I began to analyze the *Compendium* data in earnest. Each of us had separate objectives: Jack is more interested in environment and I am more inclined to evolutionary questions. We wrote programs to explore various facets of the record. Some of these were simple bookkeeping programs with the main purpose of summing, tabulating, and averaging. Others did more sophisticated statistical analyses. In some instances, we were testing hypotheses in the traditional (and stereotyped) style of research, and in others we were just exploring for interesting patterns. Although we both knew about Al Fischer's cycles, I don't thnk it ever occurred to us to use the new data base to test for them.

THE 26-MILLION-YEAR PERIODICITY

In the late spring of 1983, Jack and I were working with some purely graphical output in search of obvious patterns in the pace of extinction through time. Was extinction continuous or episodic? Was there a clear difference between mass extinctions and the normal flow of background extinction between the big events? Were the big extinctions really different from background? At this stage, we were looking at the computer output mostly as a series of pictures—looking for a gestalt that could lead us in interesting directions.

On some of the output, we could see most of the major and minor extinction events rather clearly, largely because of an image-enhancing effect of the plotting routine. The extinctions seemed to be regularly spaced in time, or at least far more regularly spaced than if they had been placed at random. Could this be Al Fischer's big cycle of 32 million years? We thought not. But the idea persisted. We even looked at the plots from across the room so the gestalt might come through more clearly, hardly a rigorously scientific procedure.

In any event, the pattern looked enough like Fischer's to warrant a closer investigation using the formal methods of statistical analysis. I won't go into the excruciating details of the analysis, but I do want to say enough to provide the flavor.

The bookkeeping programs we ran on the *Compendium* data showed where extinctions were concentrated in time. Each major pulse of extinction from the Permian onward could thus be placed on a time line. The terminal Cretaceous mass extinction was one point, the late Permian another, and so on. In the initial analysis, we identified twelve extinction events. The question was whether the spacing of these twelve was purely random or indicated a pattern.

Points placed at random on a line get us into some interesting problems, which have analogs in many aspects of human experience. The answers are often counterintuitive. Suppose, as a hypothetical experiment, that we were to draw a card from an ordinary deck of cards every morning for 250 days. If the card is a black ace, we put an X on a calendar for that day. The card is replaced and the deck shuffled for the next day. At the end of the 250 days, there will be a scattering of X's, but how will they be spaced?

On average, the X's should occur every twenty-six days, because the chances of getting a black ace are 2/52 on any given day. This number can vary considerably by chance, however.

Now, let us see what these experimental distributions really look like. This is where it gets interesting. As an illustration, I have run the experiment five times, using a computer program with a random-number generator to avoid having to wait around for 1,250 days. The results are shown below.

The "days" run from 1 at the top to 250 at the bottom, and the five "runs" of the computer program are shown in columns lettered A to E. Notice how irregular the spacing is. In column E, for example, eleven black aces were drawn,

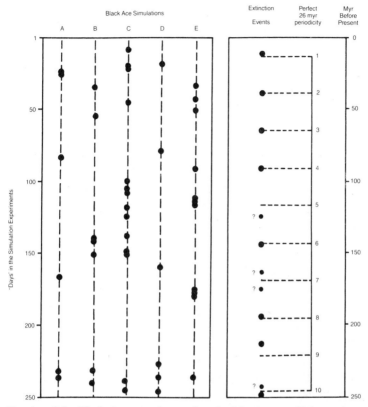

Results of the Black-Ace experiment described in the text. Using a time scale from 1 to 250 days (left side of diagram), the five columns indicated by the letters A–E show the days on which a black ace was "drawn" from a shuffled deck of cards (for five separate computer simulations). The average spacing between black aces (the small dots) should be 26 days because the chances of drawing a black ace on any given day are 1 in 26. But, as can be seen, typical results of the experiment show strongly clustered occurrences. This pattern may be compared with the more uniformly spaced extinction events from the last 250 million years of earth history (right side of diagram). It was this uniformity of spacing that led to the proposal of periodic extinction.

highly clustered: three turn up between days III and 118 and
three between 176 and 180. These clusters are balanced by
long gaps where no black aces were drawn. The results are
completely typical of points arrayed on a line at random.
Rather than a "waiting time" of about twenty-six days between black aces, we find that most gaps are smaller. A few
long gaps are balanced against a lot of short ones to produce
the average of twenty-six.

What I have just described occurs in real life, too. Rare
events such as hurricanes and floods generally show a random distribution in time. This does not mean that hurricanes
and floods are random in the sense of having no cause; it
simply means that the causes are so many and complex that
they behave randomly when taken as a group.

We often hear about 100-year floods in New Orleans or
that hurricanes hit St. Thomas on average every twenty-five
years. From the black-ace experiment, we know that the
100-year flood does *not* occur like clockwork every 100 years.
Even though the average spacing over a long period may be
100 years, one's actual experience on the ground is one of
clusters of closely-spaced "100-year" floods and long calm
periods of more than 100 years. The same is true of hurricanes. Despite conventional wisdom, St. Thomians are not
safe today merely because they had a hurricane a few years
ago.

To return to the narrative, Jack and I thought we could
see a non-random pattern in the distribution of extinction
events. That is, the events looked to be more evenly placed
in time than would be expected of a random pattern such as
I have shown with the black-ace experiment. In the illustration of the results of the black-ace experiment, I have indicated the placement of the twelve extinction events we had
identified from the *Compendium*. The 250-unit scale is con-
venient because we were working with about 250 million
years of geologic time. I have also indicated a perfectly

spaced set of 26 million-year intervals adjusted to the "best fit" position for the extinctions. This gives a prediction of the placement in time of ten extinction events if extinction occurs with perfect regularity every 26 million years.

The four small points with question marks were weak events in the original analysis, and we later showed them to be of doubtful statistical significance. If you look just at the eight points without question marks, I think you will agree that the match to the 26-million-year prediction is very good. Two of the predicted events, numbered 5 and 7, are missing unless you accept one or more of the events with question marks, and one of the predicted events, number 10, may be double.

At a purely qualitative and subjective level, what I have just shown is the guts of the periodic-extinction argument: greater regularity than would be expected by chance. But this does not settle the question by any means. The figure of 26 million years for the interval has no prior justification. It is merely a figure that best fits the observations. Furthermore, the scale of "predicted" events has been shifted until the differences between observed and predicted are as small as possible. With procedures of this kind, it is always possible to induce the picture to look better than it really is. The road to good scholarship is paved with imagined patterns. Much more elaborate testing had to be done.

I should note in passing the apparent conflict between Fischer's 32-million-year cycles and our 26-million-year figure. There is actually very little difference, because the geologic dates of the rocks of the Mesozoic and Cenozoic eras have changed considerably from the time scale Fischer was using. When our data are compared with Fischer's, it turns out that most of the extinction events are the same ones. So, if Fischer had worked with modern time scales, he would have hit on 26 million years or something close to it.

Jack and I spent most of the spring and summer of 1983

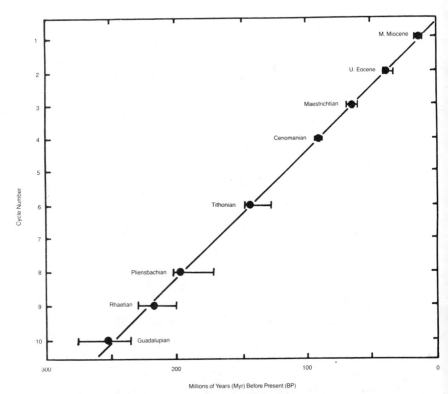

Periodic extinction events for the past 250 million years. The dots show the most probable positions in time of the eight statistically significant extinctions. The horizontal bars show the "worst case" uncertainty in the dates of the extinction events. Each of the events has been assigned a "cycle number" following the hypothesis that the events are exactly 26 million years apart. The straight line defines a perfect fit to the 26-million-year periodicity. To make this periodicity work, one must postulate that two events, numbered 5 and 7, are missing from the record: either they did not occur or they have not been found. The most recent four events (upper right) are the best dated and fit the hypothesis of periodicity almost perfectly. The K-T event of dinosaur fame is the third one back, labeled Maestrichtian for the last major subunit of the Cretaceous period.

analyzing and re-analyzing our data. It was a nerve-racking time. We were stimulated but skeptical. We wanted to avoid self-delusion. Much of the time was spent trying to kill the periodicity—that is, trying to show that it was just an accident or an artifact of our own procedures. Typically, one of us would arrive at the lab daily with a new brainwave. Perhaps the units of the geologic time scale were periodic and not the extinctions assigned to them. Or maybe the fact that we were testing all periods from 12 to 60 million years made it inevitable that one would appear to be statistically significant. Always there was a good chance that the work of several months would go down the drain.

The most important task was to be able to reject the proposition that the *appearance* of even spacing was just luck in a basically random system. What are the odds of eight or twelve randomly placed points accidently falling close to a periodic pattern of 26 million years or any other set number? In order to be publishable, the results had to reject this hypothesis of randomness with at least a 95 percent chance of being right, the conventional minimum standard.

Someone outside science may wonder why the standards for statistical testing are so harsh. Why insist on a research result being 95 percent sure? Wouldn't anything over 50 percent be "probably correct" and thus worth communicating to other scientists? In fact, the seemingly harsh standards have a good justification. Suppose we published all results in which we had better than 50 percent confidence. This might be all right for the individual result but when done repeatedly would mean that many of our published conclusions—perhaps approaching half—would be dead wrong. Even with the 95 percent standard, something like one in twenty conclusions is probably wrong. If a single conclusion is based on a chain of several inferences, each with 95 percent confidence, the final conclusion will probably be wrong if the chain is long enough. It is for this reason that many fields of science insist on having at least 99 percent confidence.

Generally our level of statistical confidence in the extinction analysis was much higher, well over 99.9 percent, but we continued to hammer at the problem. We used every standard and non-standard mathematical technique we could find or devise.

THE BERLIN CONFERENCE

Since the periodic pattern would not go away, we finally got to the point of discussion with colleagues. Along with this came seminars on the research and the beginnings of a manuscript. In May of 1983, I went to a Dahlem conference in Berlin. Dahlem conferences are wonderful experiences. For each conference, about fifty people from all over the world are invited to spend a week in West Berlin discussing a single topic or a few related topics, usually highly interdisciplinary. All expenses are paid by the city of West Berlin, the Deutsche Forschungsgemeinschaft (West German science foundation), and the Stifterverband für die Deutsche Wissenschaft (a private foundation in Berlin). An unusual feature is that there are no formal presentations, only discussion in an atmosphere of total immersion.

The Dahlem conference in May of 1983 had the working title "Earth History: How Smooth, How Spasmodic?" and the group of people in attendance was ideal for my purposes. I wanted feedback and help. If Jack and I were off base, this group could set us straight. On the other hand, I was hesitant to talk about the research too much—it was so new and raw, and possibly embarrassingly wrong. I was not concerned about giving away secrets and getting scooped, because it was unlikely that anyone else was working along the same lines, and Jack's data base in its computerized form was not generally available at that time. Also, most of us have learned that the gains that come from sharing information and getting

Some of the participants at the May, 1983, Dahlem Conference in Berlin. Included are several people who were to become prominent in the Nemesis Affair. Standing (from left) are: WALTER ALVAREZ from Berkeley, Dieter Fütterer from the Wegener Institute in Bremerhaven, Andreas Wetzel from the University of Tubingen (West Germany), BRIAN TOON from NASA who did some of the early atmospheric modeling of the effects of a large meteorite impact and who later transferred this to the nuclear-winter scenario, Kevin Padian from Berkeley, EUGENE SHOEMAKER from the U.S. Geological Survey in Flagstaff, who knows more about asteroids and meteorite craters than almost anyone, and DIGBY MCLAREN of the University of Ottawa, who, in 1970, suggested meteorite impact as a cause of mass extinction. Seated (from left) are: the author, JAN SMIT from Amsterdam, who was to do some of the most critical (and controversial) geological analyses of the K-T boundary, Tove Birkelund from Copenhagen, KEN HSÜ from Zurich, who was to do many innovative studies of mass extinction, including the development of the idea of the "Strangelove Ocean," and Jere Lipps from the University of California at Davis. (Photo: Elke Petra Thonke, Berlin)

feedback far outweigh the possible loss of priority.

Al Fischer was not at Dahlem. He had been invited and planned to go but had to cancel at the last minute because of an illness. This was unfortunate because Jack and I were most eager to have his reactions. But a number of other very appropriate people were there, including Ken Hsü from Zurich, Walter Alvarez from Berkeley, Eugene Shoemaker from the U.S. Geological Survey, Jan Smit from Amsterdam, Digby McLaren from Ottawa, and Brian Toon from NASA.

As it happened, there were no big discussions of periodic extinction at Dahlem. This was partly because I wanted to keep it low key, but mostly because all of us were consumed with interest in the impact hypothesis of the Alvarez group and the increasing controversy that surrounded it. I did spend quite a bit of time, however, talking with Gene Shoemaker. Shoemaker is awesomely knowledgable about asteroids and meteorites. Many of you have seen him sitting on the edge of Meteor Crater talking about impacts in one or another television documentary. Also, Gene has done more than anyone else to sight and record present-day asteroids in potentially Earth-crossing orbits. He has a good handle on the hundred or so authenticated impact craters on Earth.

Mostly, Gene and I centered on the intriguing possibility that meteorite impacts might be periodic themselves. If impact could cause one mass extinction, the K-T boundary event, maybe impacts were causing others—perhaps all of them! Gene was not naturally drawn to this idea, because all his work on craters has assumed that they hit at random times, but he has a very flexible mind and was intrigued. We worked over the list of craters (published in the book that had come out of the Snowbird Conference) to see if the ages fit Jack's and my extinction events. Because many of the craters are very weakly dated, we had to cull the crater list for a short list of well-dated ones. Gene could do this by eye, so familiar was he with the source data.

Gene and I didn't find much. That is, the crater ages did not look periodic and they did not match the extinction ages in any convincing way. But we had no computer at Dahlem. There was a big gap between doodling and serious analysis. Four or five months later when Gene and I were together in Chicago, we ran some simple Fourier analyses on the culled crater list to look for periodicity. Fourier analysis is a standard, quick-and-dirty method of looking statistically for regularity in time series. We found absolutely nothing.

Imagine our surprise, then, when shortly after, later in the fall, Walter Alvarez and Rich Muller at Berkeley announced that they had found a 28-million-year periodicity using almost the same crater list. I strongly suspect that Gene and I came up empty because we really didn't expect to find anything. We may have been victims of the reverse of the old saw: "I would not have seen it if I had not known it was there." Or maybe Alvarez and Muller were trapped by their own expectations.

The few journalists at Dahlem had been asked to keep a low profile. A few stories did emerge from the meeting, however. I tried quite hard to keep my periodicity results out of those stories and almost succeeded. One by Richard Fifield in the British magazine *New Scientist* mentioned the research, and even showed a sketch of me presenting the data, but his discussion was confined to one rather noncommittal paragraph.

SEPKOSKI'S FLAGSTAFF PRESENTATION

Some months after the meeting in Berlin, Jack Sepkoski was to speak at the "Dynamics of Extinction" symposium in Flagstaff, Arizona, a meeting organized primarily by university scientists in Arizona and sparked by the interest in mass extinction that had developed following the 1980 Alvarez paper.

Our confidence in periodic extinction had been growing . over the summer because the results had withstood yet more statistical testing. This was an appropriate time to go public, public only in a limited sense because the Flagstaff meeting was catering only to the geological and paleontological communities. Quite likely only a few science writers would be there to listen.

Jack presented the results of our number crunching and concluded that the 26-million-year periodicity for the extinctions of the past 250 million years was real. He also developed the notion that the clocklike spacing of extinctions could be explained more easily by calling on extraterrestrial processes than earthbound ones. There was nothing particularly profound about this choice: it simply said that regular cycles on long time scales are more common in the Solar System and Galaxy than in or on the Earth. In the environment of space, lots of bodies are circling other bodies at pretty regular rates. The Galaxy rotates completely about its axis every few hundred million years, our solar system oscillates up and down through the Galaxy in tens of millions of years, and so on.

We had no specific extraterrestrial force to suggest. After all, neither of us has any substantial knowledge of astronomy and astrophysics. In a sense, we were doing what Otto Schindewolf and Digby McLaren had done. Not being able to explain our observations by normal means, we suggested an abnormal interpretation in another discipline. But we had to stop there—hoping, of course, that the astronomers and astrophysicists would pick up the challenge. They picked it up with a vengeance.

THE PNAS MANUSCRIPT

Having gone public with our statistical results, the next logical thing was to write a manuscript for publication. So, Jack and I put together a very brief account, including the rather loose extraterrestrial suggestion, and sent it off to the National Academy of Sciences for publication in its *Proceedings*. The paper was submitted in October of 1983 and published the following February. Because the *Proceedings of the National Academy of Sciences (PNAS)* has strict page limits, our report was only five printed pages. It was densely packed and devoted mostly to very technical details of our statistical procedures. This paper has been cited and talked about many times, but I suspect that relatively few people have actually read it. This is par in an environment where hundreds of papers are published every week.

Why did we choose *PNAS? PNAS* offers several things: quick publication, a very large circulation here and abroad, and lack of peer review for papers authored by members. We wanted quick publication, although there was no compelling reason for any hurry. As far as we knew, nobody was working on the same problem, so we were not likely to be scooped, even assuming there was anything important enough for scooping. On the other hand, quite a number of people knew about the work by now and it was just conceivable that someone might dust off old research to develop the same case. I do not mean to suggest malice. Stealing the ideas of others is vanishingly rare in science. But all of us have a lot of trouble separating our own ideas from those we have picked up listening to other people. For this reason, the same research ideas have a tendency to pop up in several places at once. Also, I think Jack and I felt we had something exciting and we were impatient to get it in print. After all, research does not exist, in some sense at least, until published.

PNAS's lack of peer review was also a factor. Most substantial scientific journals send all incoming manuscripts to two or more people in the discipline. A paper is accepted or rejected primarily on the basis of these reviews. It is only practical to have two or three reviewers for each manuscript, because otherwise most scientists would spend more time reviewing other people's manuscripts than doing original research.

Having such a small number of reviewers for each manuscript makes the whole game very chancy. Bad papers are often accepted because they have gone only to bland or ill-informed reviewers, and good papers may be rejected because one or two reviewers were hypercritical, nasty, or ignorant of the research. In spite of these problems, the peer-review system is generally accepted as important. It helps to maintain a journal's quality, and it provides a reasonably fair and objective evaluation of people and their research. When a research scientist is coming up for promotion or being considered for a new position, one of the first questions always asked is: How many papers does he or she have in peer-reviewed journals?

Not surprisingly, journals that are not peer reviewed are often looked upon with some suspicion and as a sort of gray literature. Papers published in such journals carry less weight, and it is even possible to hear people criticize a piece of research just because it did not appear in a peer-reviewed journal.

PNAS is published by the National Academy ostensibly as an outlet for its members. The Academy is a self-perpetuating "club" consisting of about 1500 people whom the group itself considers to be the top scholars in the country. It was created by the U.S. Congress for the purpose of advising the government on scientific matters, and although it serves this function well, it also serves as a sort of "honor roll" of American science. Most other developed countries have comparable organizations. In some countries, such as the Soviet Union, academy membership provides substantial personal clout and privilege. In others, including the United States, membership opens a few doors and is nice for one's ego, but little more. Perhaps this is because everyone realizes that many outside the Academy are just as good or better than those inside. Since I was elected some years ago, I have experienced two "firsts" for my career: a research paper rejected by a peer-reviewed journal and a National Science Foundation proposal declined!

As a member of the Academy publishing in *PNAS*, I was not subject to peer review. I suppose there may have been some insecurity lurking in Jack's and my decision to send our paper to *PNAS*, but I would rather say we wanted to get the work into general circulation without going through the extra time and hassle of formal review. Of the many criticisms of our work that have been published since, only one or two have chastised us for avoiding the review process. In any event, periodic extinction was now fully out in the open and ready to be embraced or destroyed by the international scientific community.

8: *Nemesis Is Born*

ASTROPHYSICS AND PALEONTOLOGY

AT Flagstaff, Jack had thrown out a challenge to astronomers and astrophysicists to explain the 26-million-year periodicity. Surprisingly, they responded. Normally, such challenges go unheeded, I suppose because communication between disciplines is not very good. Each field is a highly complex culture pursuing interesting problems of its own. Furthermore, none of us has the breadth of training necessary to appreciate questions posed in most other fields—especially if we are talking about fields as distant and different as paleontology and astrophysics. Astrophysics has a reputation of being the hardest of the hard sciences: to many observers, it is higher in the pecking order even than high-energy physics. Paleontology, on the other hand, has a sort of nineteenth-century flavor: naturalists and dedicated amateurs like Louis Leakey devoting their lives to finding new fossil remains.

Fortunately, the astrophysicists heard about periodic extinction in a form they could understand. Within a few weeks of Jack's Flagstaff presentation, three good treatments of the extinction research appeared in the scientific and popular press. Roger Lewin of *Science* wrote it up as part of a general report on the Flagstaff meeting. Cheryl Simon of *Science News* followed with an excellent account based on Lewin's piece and some interviews. And George Alexander of the *Los Angeles Times* wrote a clear and authoritative article. These three writers, among the best science journalists, wrote interesting and technically accurate stories. Astrophysicists were so intrigued that a number of them started serious work on the problem. This ultimately produced Nemesis.

It should be remembered that mass extinction was on many people's minds in the late summer of 1983. The Alvarez proposal of mass extinction by meteorite impact was being debated and the science journalists were following the story. Less important but still significant was the increasing public and scientific interest in contemporary, man-made extinction in tropical rain forests. I am confident that no journalist, with the possible exception of Roger Lewin, would have covered the Flagstaff meeting except for the continuing interest in extinction. And if Jack's presentation had not been covered by the press, there is an excellent chance that astrophysicists would never have known about our challenge to them.

There is an amusing side light on the positions of astrophysics and paleontology as distant parts of the spectrum of science. Paleontology is part of geology in many respects, and geology has a curious moral authority over astrophysics. This stems from two or three instances where astrophysics has been embarrassed by discoveries in geology. The best example is that strange interval, I think in the 1950s, when the Earth was said to be older than the universe that contained it. Each of the two branches of science has its own ways of estimating the timing of events in the distant past. For a short period, astrophysicists estimated the universe to be about three billion years old and geologists saw the Earth as being somewhat younger. This was fine, because the Earth cannot be older than the universe. But then some new meth-

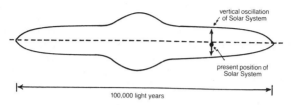

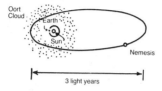

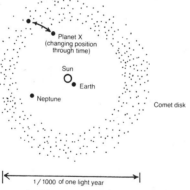

The three principal astronomical hypotheses proposed to explain periodic extinction: oscillation through the galactic plane, the companion star (Nemesis), and Planet X. *Top:* edgewise view of the Milky Way galaxy showing the present position of our solar system near the galactic plane. The Solar System crosses the galactic plane every 31 to 33 million years. *Middle:* proposed orbit of Nemesis in relation the Solar System. Nemesis is said to pass through the Oort Cloud of comets every 26 million years. *Bottom:* our solar system showing the comet disk in relation to the changing position of Planet X in its eccentric and changing orbit. It should be emphasized that none of the diagrams is drawn to scale: if this were attempted in the top sketch, for example, the Solar System would be invisibly small.

ods were employed by geochemists to gauge the age of the
Earth. They came up with the present figure of about 4.5
billion years. It did not take long for astrophysics to get its
act together. Soon, the universe was estimated to be 17 to 20
billion years old—a comfortable margin.

I don't know for sure whether the age-of-the-Earth-and-
universe problem really had any influence on the way astrophysics received the arguments about periodic extinction.
The anecdote about the age of the universe does emphasize,
however, that there can be strange dynamics operating between scientific disciplines.

In any event, a number of very bright astrophysicists were
sufficiently intrigued by the claim that extinctions occur
every 26 million years to spend research time and energy
trying to find a cosmic explanation. Keep in mind that this
interest developed before our work had been tested and evaluated by our own colleagues in paleontology. All the astrophysicists had were press reports of research being done by
two Chicago paleontologists they had probably never heard
of.

NATURE: APRIL 19, 1984

The April 19, 1984, issue of the British journal *Nature* had a
color photograph of Meteor Crater, Arizona, on its cover
and the words *mass extinctions.* The issue contained not one
but five scientific papers and two editorials on the subject of
periodic extinction in the geologic record. The five research
papers were arranged in the order of their receipt by the
editors of *Nature,* as follows:

Received November 15, 1983: M. R. Rampino and R. B.
Stothers recalculating the statistics of extinction, coming up
with 30 million instead of 26 million years, reporting a significant periodicity of 31 million years for impact craters, and

proposing that the Sun's motion perpendicular to the galactic plane is responsible for both. *Received November 16, 1983:* R. D. Schwartz and P. B. James interpreting the 26-million-year extinction periodicity as being caused by the Sun's motion perpendicular to the galactic plane. *Received January 3, 1984:* D. P. Whitmire and A. A. Jackson IV interpreting the 26-million-year extinction periodicity as being caused by an as-yet-unseen companion star. *Received January 3, 1984:* M. Davis, P. Hut, and R.A. Muller interpreting the 26-million-year extinction periodicity as being caused by an as-yet-unseen companion star. They named the unseen companion "Nemesis." *Received January 30, 1984:* W. Alvarez & R. A. Muller reporting that impact craters occur periodically every 28 million years.

In these five papers, there was a largely confirmatory recalculation of the extinction statistics, two analyses of impact crater ages showing periodicities close to that for extinction, and two different astrophysical interpretations (galactic motion and the companion star). The two galactic-motion papers arrived at *Nature* on consecutive days in November and the two papers on the companion star on a single day in January.

As you will have noticed, the five papers were submitted for publication before Jack's and my paper was even published. John Maddox, *Nature*'s editor, included some mild wrist slapping of Jack and me in his editorial in the same issue. He noted that had the five papers not needed revisions, they might actually have been published before the appearance of our *PNAS* paper on which they were based. He felt that this raised the not-uncommon problem that circulation of preprints to an inner circle of colleagues cuts out equally worthy scientists who are not part of the network. Maddox

had the following comment concerning circulation of preprints of forthcoming articles:

> This practice, usually intended as a courtesy to colleagues else-
> where . . . can also be thoroughly unhelpful to other people. The
> most obvious complaint against the system is that it is dis-
> criminatory, excluding from those in the know people who hap-
> pen not to be on the authors' mailing list.

I think Dr. Maddox raised an issue that should concern
all practicing research scientists. In the present case, however, he was not aware of all the factors. He realized neither
the effects of the journalists' accounts of the Flagstaff meeting nor that networking was of little or no importance in this
instance. Jack and I had sent a preprint of our *PNAS* paper
to the Alvarez group in Berkeley because of the obvious
connections with their research on mass extinction. Beyond
that, we sent perhaps a dozen unsolicited preprints to paleontological and geological colleagues. The astrophysicists got
preprints because they heard about the work and requested
copies of the manuscript.

The rest of the Maddox editorial was a thoughtful and
well-reasoned discussion of the whole business. It did not
judge the new research, although it took pains to point out
problems with both the paleontological and astrophysical
analyses. Despite the somewhat negative flavor, the Maddox
editorial was a good introduction to the suite of papers. This
was followed by a much more technical commentary by
Anthony Hallam of the University of Birmingham. He expressed great doubt that Jack and I had proven the case for
periodicity.

Let me now give some more detail on the astrophysical
explanations proposed in the April 19, 1984, issue of *Nature*.

THE SUN'S MOTION IN THE GALAXY

The Milky Way Galaxy, of which we are a part, is a complex of moving stars. The Galaxy is disk-shaped. As it rotates in space, our Sun and its planets, asteroids, and comets move slowly up and down across the galactic plane. This motion has been known for a long time and the astronomical literature contains estimates of the length of time required for the Solar system to complete a full cycle of movement—from below the galactic plane to above the plane and back to an original position below the plane. This cycle time is thought to be somewhere between 62 and 67 million years: several figures in this range are quoted by the two *Nature* papers. The Sun crosses the galactic plane twice in each complete oscillation, and therefore the crossings of the plane come every 31 to 33 million years. (The estimates are not precise enough to worry about the fact that 67/2 is 33.5 rather than 33.)

It is generally thought that as we near the plane of the galaxy, various aspects of our cosmic environment change. We have a greater likelihood of encountering interstellar clouds of gas and dust, and the levels of certain kinds of radiation may increase. Both the Rampino/Stothers and the Schwartz/James papers in *Nature* suggested that effects of approaching the galactic plane might be producing biological effects on Earth. Rampino and Stothers would have close encounters with interstellar clouds perturbing Oort Cloud comets and thereby increasing the chances of comet impact on Earth. Schwartz and James suggested that the increases in cosmic and other radiation would affect Earth climates, which in turn could have caused the 26-million-year extinctions.

It should be kept in mind that the vertical oscillation of the Solar system cannot actually be seen, and there is considerable uncertainty about the nature of the environment near the plane of the Galaxy. To a paleontologist, astrophysicists have an appalling lack of observational data. Remember from Chapter 1 that the Oort Cloud of comets (used by Rampino and Stothers) has never been seen. But the astrophysicists seem to know what they are doing.

The galactic-plane explanations for the 26-million-year extinctions have a big advantage over the other explanations because they have a number from the geological record (26 or 30, depending on whose statistical analysis you use) that is very close to an independently derived number (31 or 33) from studies of galactic motion. There is a strange problem with the galactic-motion idea, however. Our Sun is currently very close to the galactic plane, yet the most recent extinction Jack and I recognized is the one 11 or 12 million years ago. The extinctions thus suggest that we are now about midway between two events, which should put us near the maximum distance from the galactic plane. Something is wrong! Not surprisingly, the proponents of the galactic-plane explanation have wiggled out of the paradox, while the advocates of a companion-star explanation find the mismatch in timing to be a devastating problem.

THE COMPANION STAR

The two papers proposing the companion star were remarkably similar. Both postulated a small solar companion on a highly eccentric (non-circular) orbit—an orbit that carries the companion through the Oort Cloud once per revolution about the Sun. Accidental disturbance of comet orbits in the Oort Cloud then produces a comet shower on Earth and the comet impacts, causing mass extinction.

Both papers concluded that the companion star must be very small (probably considerably less than a tenth the mass of the Sun) and positioned now about two light years from Earth. Many other characteristics of the companion also jibed in the two papers. This is not surprising. Both teams of astrophysicists were trying to satisfy the same constraints, including the need for a 26-million-year orbital period and the need for the companion to pass through the Oort Cloud once per revolution but to remain in the Oort Cloud only a short time. All in all, they wrote a credible scenario that would satisfy the requirements implied by accepting Jack's and my extinction periodicity.

If there is a companion star, why have we not seen it? This is a natural question and one addressed in both papers. At a distance of two light years, the companion would be by far the closest star to the Earth, about half the distance to the next closest (Proxima Centauri). It turns out that there could easily be a star at very close range that has been overlooked. Of all the stars visible in the sky, surprisingly few have been catalogued, and of these only a small number have been studied fully enough to establish their distance from the Earth. A very dim star close to Earth can easily be confused with a bright star farther out unless measurements of proper motion and parallax are made. These measurements have never been made systematically. It is by no means impossible that a little star is "hiding" very close to us.

Another obvious problem is whether the orbit of the proposed companion star would be stable over long periods of geologic time. Among other factors, miscellaneous stars in the Galaxy may by chance pass close to us and thereby deflect the orbit of the companion star by gravitational effects. Such close encounters would be exceedingly rare, but geologic time is vast enough that this possibility becomes real in the realm of tens and hundreds of million years.

The companion-star idea is an interesting example of a scientific style. The whole argument is *ad hoc*. In formulating the model, the 26-million-year periodicity of extinction is accepted. This acceptance creates the need for an orbit in the companion which will, when tied to the Oort Cloud, do the job. The result is testable—at least to a degree. The stability of the orbit can be analyzed. The whole idea would be falsified if the orbit proved impossibly unstable.

More directly, a search can be mounted for the star itself. Richard A. Muller, an author of the second *Nature* paper proposing the companion star, now searches actively for it by optical telescope. Using a highly computerized system, he is starting with a list of likely candidates from established star catalogs and observing each over a period of time. If the star catalogs are exhausted without success, Muller will move to more general sky searches in both the northern and southern hemispheres. If he finds the star in its predicted orbit, the model will have been confirmed. If he fails, we will not have learned much. The star may still be out there but unrecognized.

NEMESIS VERSUS SIVA

As I mentioned, the Davis, Hut, and Muller paper in *Nature* dubbed the putative companion star "Nemesis" and the name has stuck. Their christening statement went as follows:

> If and when the companion is found, we suggest it be named Nemesis, after the Greek goddess who relentlessly persecutes the excessively rich, proud, and powerful. We worry that if the companion is not found, this paper will be our nemesis.

Actually, the three authors suggested several other names in a footnote to their original manuscript, but the editors of *Nature,* in their wisdom, eliminated all but Nemesis.

A few months later, the name was challenged in a delightful and shrewd essay by Stephen Jay Gould in *Natural History.* His criticism was in the form of an open letter to Davis, Hut, and Alvarez and went as follows (in part):

If Thalia, the goddess of good cheer, smiles upon you and you find the Sun's companion star, please do not name it (as you plan) for her colleague Nemesis. Nemesis is the personification of righteous anger. She attacks the vain or the powerful, and she works for definite cause . . . She represents everything that our new view of mass extinction is struggling to replace—predictable, deterministic causes afflicting those who deserve it.

Gould's last point is quite serious and reflects one of the main themes of his essay: that mass extinction by comet impact may not be a fair game, with the better-adapted organisms surviving and the others perishing. To amplify this idea, Gould urged that the companion star be called "Siva" after the Hindu god of destruction:

Unlike Nemesis, Siva does not attack specific targets for cause or for punishment. Instead, his placid face records the absolute tranquility and serenity of a neutral process, directed toward no one . . .

It is curious indeed that there should be such scholarly debate over the name of a heavenly body that has never been seen and may not exist. Perhaps it is fitting that all the candidates are gods and goddesses.

Whether Nemesis exists or not, she has taken on a life of her own in scientific theory and debate. I often hear astrophysicists say things like "The Nemesis orbit requires . . ." or "Your interpretation can't be correct because it does not fit with Nemesis." The astrophysicists know, of course, that Nemesis neither has been seen nor is required by theory to exist. But this is the way we move forward in research. The companion-star hypothesis makes certain predictions and the hypothesis is considered viable until proven otherwise. Any field of scholarship wishing to extend its horizons has to use this *modus operandi.* The system works well as long as participants remember which theories have been confirmed and which have not.

You will remember that when Gene Shoemaker and I did some rough-and-ready testing of the impact-crater list in the fall of 1983 to see if cratering on Earth has been periodic, we drew a blank. A surprising ingredient of two of the April 19 *Nature* papers was statistical analysis claiming periodicity in the crater ages. Rampino and Stothers found a 31-million-year period and Alvarez and Muller found a 28-million-year period.

I remember vividly the day Alvarez and Muller found their 28-million-year periodicity. Jack Sepkoski and I got word that the Berkeley people wanted to have a conference call about extinction. The telephone setup was worked out and five of us started talking: Jack and I in Chicago and Walter Alvarez, Rich Muller, and Luis Alvarez in Berkeley. They told us excitedly about the cratering periodicity and their enthusiasm for it as confirming both periodic extinction and the basic idea of extinction by large-body impact (for more than just the terminal Cretaceous event).

Until this time, the Alvarez group (especially Alvarez senior) had been somewhat skeptical of our periodic-extinction analysis. Part of this doubt, I think, was because they subscribed to the conventional wisdom that meteorites hit the Earth at random times. It is also conceivable that they saw the periodicity idea as antagonistic to the meteorite explanation for the Cretaceous mass extinction: the mechanism for periodic mass extinctions might involve some driving force other than meteorite impact. It should also be said that Luis Alvarez had some excellent criticisms and suggestions to make about our statistical analysis of extinctions. He did not find our original work convincing.

In any event, periodic cratering made it reasonable to pull together all the disparate elements of the extinction problem: cratering, periodic extinction, and the K-T event. The only problem was that Jack and I were skeptical of the Berkeley cratering analysis. The number of well-dated craters is very small, and I, at least, was still influenced by the failure Gene Shoemaker and I had encountered in our search for periodic cratering.

Our skepticism faded within a few hours of the conference call, however, after Jack and I had done our own cratering analysis, using the same computer programs we had used for the extinction analysis. Using several different ways of selecting craters, we got exactly the same result as our Berkeley colleagues: 28 million years. This was especially striking because our methods were totally different.

In spite of this wonderful conformity, some remain skeptical to the point of denying that the craters show any pattern. And, in truth, the crater ages are very uncertain.

PLANET X

Yet another explanation for the extinction periodicity has surfaced. D. P. Whitmire and J. J. Matese published a paper in *Nature* in January, 1985, suggesting that the comet showers could also be produced by an unseen tenth planet, Planet X, lying beyond the orbit of Pluto. This is the same Daniel Whitmire who was the senior author of one of the companion-star papers in the April 19, 1984, *Nature*. The idea of a missing planet in our solar system has been kicking around for a long time because of possible (but debatable) discrepancies between observed and predicted motions of the outer planets. So, Planet X has some prior appeal.

Although the Planet-X idea uses comet showers to explain the extinctions on Earth, the source of the comets is very different from that of the Nemesis scenario. The prominent astronomer Gerard Kuiper, among others, postulated the existence of a disk of comets beyond the orbit of Neptune. The comet disk, assuming it exists, is nowhere near the Oort Cloud; the disk is about 35 astronomical units from the Sun and the main Oort Cloud is 20,000 to 40,000 astronomical units out. An astronomical unit (AU) is equivalent to the average distance between the Earth and the Sun (about 150 million kilometers, or 93 million miles). To some astronomers, Kuiper's comet disk is merely the inner edge of the Oort Cloud.

The scenario proposed by Whitmire and Matese is a clever one and fairly complex. I won't attempt to describe it except to say that it involves the regular changes (precession) of the Planet-X orbit such that the planet sweeps through the comet disk every 28 million years, perturbing comet orbits and producing a comet shower on Earth.

The hypothesis is *ad hoc,* of course, just as the companion-star hypothesis is, and the case has many of the same problems of proof and falsifiability. Whitmire and Matese find their new idea more appealing because it builds on the prior problem of discrepancies in the orbits of the outer planets, and because they estimate that the orbit of Planet X would be more stable over geologic time than the orbit of Nemesis. They also argue that their Planet-X scenario explains some features of the Solar System quite unrelated to the extinction-periodicity problem, such as the persistence of short-period comets.

But why have we not seen Planet X if it is there? Whitmire and Matese suggest that the answer may be that the extra planet is at an unexpected inclination to the plane of the other planets and that people have been looking in the wrong place! They further suggest that the ongoing infrared survey (IRAS) could be used to detect Planet X.

So, we have an ample collection of astrophysical explanations for periodic extinction. None has been proven right or wrong, although a number of critical studies have argued vehemently against one or another of the explanations. The Nemesis idea has gotten far more public attention than the others, perhaps owing to the name. As of late 1985, I am not hearing compelling arguments on any side. Nemesis and Planet X have some natural appeal because they are amenable to actually being discovered and really verified—if they are there—whereas no simple proof is available for the galactic-plane idea. But this is no reason to prefer Nemesis or Planet X over the other proposal. As a paleontologist, I am enjoying the whole affair and hoping that somebody is correct.

9: *Mounting Controversy*

THE Nemesis idea has caused a storm of argument and debate. The fact of such debate is not surprising, because science is basically an adversarial process. The debates surrounding Nemesis, however, and its associated claims of extinction by impact and of periodic extinction, have had greater tempo and volume than usual. Thus, they illustrate the phenomenon of controversy more clearly than other examples.

CONTROVERSY IN SCIENCE

Scientific debate takes several forms and occurs in more than one arena. The formal (or official) arena is the scientific journal. A controversial idea or set of conclusions usually stimulates people to do additional research on a problem. Once the new research is done, the results are written up and put through the peer-review process, and an article may be published. The scientific community can then evaluate the new work in comparison with its predecessor. Except in open-and-shut cases, the process may have to be executed several times before the originally controversial idea is accepted or rejected.

Although this scenario is routine, there are many variants, and often the formal system is overshadowed by simultaneous events. The most common variant is that a controversial paper will be rebutted explicitly in a later issue of the same journal. *Nature* has a special section under the heading "Matters Arising." *Science* publishes what it calls "Technical Comments." And so on. The form of these exchanges is quite fixed: the rebuttal piece is sent by the editors to the author of the original paper for his or her response. Then, the rebuttal and response are published together. Thus, the original author has the last word of the moment.

When a truly controversial idea or conclusion comes up, the whole neat process tends to break down. This certainly happened in the case of the Alvarez theory of dinosaur extinction and even more so with periodic extinction and Nemesis. A hot topic short-circuits the system. Journal articles require from two months to two years to process and publish; the average is perhaps ten to twelve months. I am told that in certain fast-moving parts of high-energy physics and molecular biology, questions are debated and settled long before the first paper is published. The papers are still published, of course, for their archival importance. Also, publication is the life blood of the curriculum vitae.

A great many controversies are waged as much by rumor as by formal means. "Have you heard that the shocked quartz at the K-T boundary could easily have come from diamond pipes?" "Did you know that there is more iridium in some coal than Alvarez ever found in the K-T boundary clay?" "I understand that their statistics do not hold up." "The whole Oort-Cloud idea is under fire." "I understand that Joe Jones at Caltech has found a fatal flaw." And on and on. Some rumors of this kind are true, of course. No law in science says hearsay has to be wrong. And informal reporting of things read or heard is certainly a valid way of communicating. But this kind of communication can be extremely damaging to the truth, just as it is in human affairs in general. In fact, rumors are probably a lesser problem in science than in most other fields: politics, for example. Science, at least, has some reasonably good mechanisms (such as peer-reviewed publication) to squelch unfounded rumors.

Rumors have been flying thick and fast in the fields cognate to the Nemesis Affair, especially so since so many participants are working outside their areas of primary training. I was trained at Harvard as a geologist, yet I did not realize that diamond pipes don't have enough quartz of any kind to worry about. And I had not heard of iridium until five years ago.

Controversy in science is also worked out in a variety of other media. Most active scientists spend much of every year presenting their work—and their ideas about the work of others—at formal meetings and in invited lectures and seminars at their own and other institutions. In the go-go days of the 1960s, it was often said facetiously that university presidents and deans gauged success by the numbers of faculty members on airplanes at any given instant. Things have calmed a bit, but speaking before groups is still a primary mode of scientific communication. And the lectures are almost never published.

The press, both scientific and lay, also plays a role. It has the effect of speeding up the flow of information and opinion. In some cases, it plays a role in the course of the research itself.

A special problem involving the press deserves mention here. Most journals have a policy of not publishing research results that the authors have already published elsewhere. This is a completely reasonable policy that can get out of hand. What constitutes being published elsewhere? George Alexander's account in the *Los Angeles Times* of Jack's Flagstaff presentation? Or review in a scientific paper of a preprint of someone else's work? There are no consistent policies from journal to journal, but all attempt to control prior publication to some extent. The great danger is that a journal will refuse a paper because mere news of it has appeared elsewhere. This could lead to the absurd situation wherein coverage by the press invalidates a research result.

Finally, a word about the origins of scientific controversy. Most routes to controversy are clear, obvious, and no different from those operating throughout our society. New ideas breed disagreement. In fact, disciplines that lack ferment are probably in trouble: either they have no creative people or the conventional dogma is so strong that change is impossible.

In my corner of science, there are a few other important stimulants of controversy. A scientific paper is generally expected to present something new. This may take the form of new data amounting to new knowledge, or the paper may support a particular concept or theory. It may also launch an entirely new concept or theory. Among these possibilities, new concepts are most valued and simple additions to factual knowledge least valued. At the lowest end is the "negative result" where someone has searched for something, or tried an experiment, and failed. Negative results, even those with far-reaching implications, may be impossible to get published. So, with a premium on newness, there is a natural tendency to exaggerate the novelty or distinctiveness of one's research results, and this can produce controversy where none is justified.

The rhetorical form of scientific articles also has a strange influence. Most articles consider two or more alternative conclusions or interpretations of data and then come down strongly in favor of one. It is rare indeed to read an article ending with anything other than a firm conclusion. One does not often encounter frank admissions of uncertainty, although statements like "the data suggest" are often used as mild caveats. I submit that the tradition—and it is only a tradition—of firm conclusions creates controversy where none is merited.

PERIODIC EXTINCTION UNDER FIRE

Jack Sepkoski and I published our *PNAS* paper in February of 1984, although our Dahlem and Flagstaff presentations plus word-of-mouth had spread widely well before. The reactions varied but tended to sort out along certain lines. The press in those early days generally accepted and promoted periodicity. This is not surprising, because journalists are in the business of reporting new discoveries, and to many journalists, a fact is a fact. Also, the periodicity story probably had more appeal to the general public if it could be assumed correct. Many an early newspaper, magazine, and television report included statements from other scientists, especially paleontologists. Some of these were critical of our conclusions, but these criticisms tended to be downplayed. Furthermore, few of our colleagues knew enough about the work then to make substantive remarks.

Scientists outside geology and paleontology generally accepted our conclusions either on the basis of press accounts or from perusal of the *PNAS* paper. The astrophysicists who came up with the Nemesis and other explanations certainly bought our reasoning. There was no reason that they should not have.

But in 1985, increasing numbers "in the trade" looked seriously at the periodicity problem. If major extinctions were indeed on a precise time schedule, and if the driving force was extraterrestrial, views of the history of life would have to change, and change drastically. Mass extinctions could no longer be looked upon as culminations of long, complex interactions among organisms and between organisms and their environments. The turmoil everyone recognized in the history of life could not be understood as an extension of processes that can be studied in present-day environments. Most important, the new hypotheses claimed that the biology of the Earth is, in the long run, influenced strongly by its cosmic environment. Lyellian uniformitarianism might thus have to yield to Cuvierian catastrophism.

The considered reaction among paleontologists was largely negative, and an arsenal of substantive criticisms developed, which included the following major points:

1. Raup and Sepkoski did not use a standard definition of mass extinction. They would have gotten different results if they had used different measures of extinction intensity.

2. The fossil record is too incompletely known for broad and valid statistical analysis. The number of extinctions in a given interval of geologic time is too much a function of how many people have studied that interval and with what scientific philosophies.

3. The taxonomy of most fossil groups is too messy to allow use of catalogs of families and their time ranges.

4. The uncertainty in geologic dating undermines any attempt to track the history of life with enough precision to find 26-million-year cycles even if such cycles are present. No two geologic time scales agree.

5. The appearance of a 26-million-year periodicity may just be a result of the uncertainties in classification and dating of fossils.

6. Raup and Sepkoski used a culled sample of only 567 families and this may have led to the 26-million-year result.

7. In the analysis, family extinctions were assigned to stratigraphic intervals, averaging a little more than six million years each. If extinction intensity is behaving like a simple random walk, one would expect a peak every four intervals. This explains the apparent regularity in the spacing of events.

8. The appearance of regularity in the spacing of extinc-

tions is due to each extinction's being caused by different and independent forces.

9. Extinctions are complex events controlled by many independent factors. A search for simple causes is futile.

10. There is ample evidence that long-term changes in sea level are the major cause of mass extinction.

11. There is ample evidence that long-term changes in climate are the major cause of mass extinction.

I will not attempt here to discuss or refute these arguments point by point. I doubt I could be objective anyway. There are some general common denominators, however, that are worth discussing.

As you will notice, several of the criticisms relate to the generally "noisy" or uncertain nature of the data Jack and I were using. The basic argument is: If there is uncertainty in the observational data, any conclusions based on them will be uncertain. This is true in some things, but not with the kind of statistical testing Jack and I were doing. You will remember in my discussion of random spacing of events (the black-ace experiment in Chapter 7) I said that the main job in the periodicity analysis was to be able to reject random spacing. Suppose, hypothetically, that we had a perfectly periodic record: events precisely every 26 million years. In such a case, rejection of the random hypothesis would be obvious and straightforward. Now suppose we "trash" the record by adding uncertainty. We could do this, for example, by moving events forward or back in time by randomly chosen amounts. As we degrade the record in this way, the periodic signal will become weaker and weaker and a point will be reached where we can no longer distinguish the record from a purely random one. Thus—and this is the important point—by adding uncertainty to the record, we have moved it away from a simple, periodic pattern and toward randomness. From this, the fact that Jack and I included uncertain data just makes the tests for randomness more conservative. To exaggerate only a little, if periodicity shows through in spite of uncertain taxonomy and geologic dating, it must be there!

Another general problem relates to concepts of randomness. Several of the criticisms I have listed fall into the trap (or common misconception) I discussed in Chapter 7 when talking about spacing of hurricanes and 100-year floods. Random processes may produce a predictable *average* spacing of events but they do not produce *even* spacing.

The last two of the listed criticisms invoke changes in sea level or climate to explain biological extinction. This is fine and worth debating. Many paleontologists are saying, in effect, "We have good explanations for the big extinctions, so who needs periodicity and its Nemesis-type interpretations?" Suffice to say that I don't find the sea-level and climate arguments compelling. But these arguments are made by some geologists who are far more experienced than I. Also, there is the tantalizing possibility that everybody can be correct! It is just possible that the periodicity of extinction reflects periodic showers of comets and that the impacts produce long-term changes in climate and/or sea level. The timing of the changes is a problem, but the possibility should not be discarded out of hand.

How many of the objections raised by paleontologists are simply kneejerk defense of conventional wisdom? This is surely an element, but I am not sure how important.

As I noted earlier, most paleontologists have reacted negatively to periodicity. There have been some exceptions, including some excellent scholars (Steve Gould, for one). Among geologists in general, the reactions have been mixed. Some, like Brochwicz-Lewinski in Warsaw and Ken Hsü in Zurich, have plunged into the problem to look for other evidence. This has taken the form of checking Jack's and my extinction events for iridium and other indicators of large-body impact. If it could be shown, for example, that the extinction events we identified are consistently associated with impact, and if the intervening intervals do not have impacts, the game would be over. This has not happened yet, but it is clearly the direction to go. After all, Jack and I based our conclusions on statistical inference and the conclusions remain inference until they can be confirmed or denied by other kinds of evidence. The jury is still out—and probably somewhat puzzled.

NEMESIS UNDER FIRE

Nemesis and her sister explanations—motion of the Sun in the Galaxy and Planet X—have all been subjected to scrutiny and debate. Aside from some put-downs having to do with the *ad hoc* nature of the scenarios, the arguments have revolved around the physics of the problem and questions of whether the explanations are viable in a purely technical sense.

A major issue with both companion-star proposals is whether the postulated orbit would be stable over the long periods of time required. A companion star so small and far from the Sun could easily be deflected by passing stars and other galactic junk. Because such encounters are by their nature unpredictable, the argument must be statistical. Piet Hut at Princeton (one author of the original Nemesis paper) has done extensive computer simulations with randomly passing stars to estimate the average life expectancy of a companion star in its presumed orbit. The results satisfy some people but not others. There is a likelihood of "wobble" in the orbit, as well as the possibility that the companion would be thrown out of its orbit altogether.

One rather curious objection to the Nemesis idea is that a wobble in the orbit would cause an average of 10 percent variation in the length of the period—variation that is not seen in the spacing of extinctions in the fossil record. So, it is argued that Jack's and my periodicity is too perfect to be explained by Nemesis! This is ironic in view of the typical arguments made by paleontologists about the fuzziness of our periodicity.

I am not close enough to astronomy and astrophysics to judge the arguments that fly back and forth. My impression, however, is that the scientific community as a whole finds the companion-star idea technically possible but not very likely. There is enough support for it, to bring a number of research groups, in addition to Rich Muller's, into the search. The possibility of a Nobel Prize at the end of that road may be playing a role. (One nice thing about paleontology is that it has no Nobel possibilities.)

I can say virtually nothing about how the Planet-X proposal has fared. It has not been an active issue in major international journals. But I suspect it will continue to lurk if for no other reason than that it renews a longstanding problem in our solar system.

The scenario based on the Sun's motion up and down through the Galaxy has been argued vehemently, however. Remember that this idea has something going for it that Nemesis lacks: the conventional estimate of 31 to 33 million years for the interval between crossings of the galactic plane. The main problem is whether the giant molecular clouds and other matter are sufficiently concentrated near the plane of the Galaxy to produce the pronounced difference between intensities of mass extinction and background extinction required by the model. Also, it has been argued that the extinctions are too perfectly spaced to be explained by the highly irregular placement of material near the galactic plane. And there is the argument that if the galactic explanation were correct, we should be in an extinction event now rather than about halfway between two of them.

IMPACT CRATERS

The ages of impact craters on earth and their possible periodicity are important elements in the whole affair. On the
one hand, the case for the cratering periodicity looks strong.
It was discovered independently at Berkeley (by Walter Alvarez and Rich Muller) and at NASA (by Mike Rampino
and Dick Stothers). And Jack and I found the same thing
with different techniques after the Berkeley group had encouraged us with their findings. Also, when one combines the
evidence for cratering periodicity with that for impact as a
cause for extinction, a very credible whole is produced. It all
ties together—and this is important in the survival of a scientific hypothesis.

On the other hand, the statistical analyses of cratering
have been attacked with many of the same arguments that
have been used against the extinction work. It is certainly
true that the dating of craters is dreadful. So dreadful, in fact,
as to push my earlier arguments about the effect of uncertain
or noisy data a bit far. Nevertheless, I think my early argument holds. No matter how messy the data, if a hypothesis
of randomness can be rejected, it is not the messiness of the
data that is causing the rejection—unless, of course, there is
conscious or unconscious cheating in selection of the data to
be analyzed. Conscious cheating is easy to avoid but unconscious cheating is not.

Naturally, much has been made of the near match between
the cratering and extinction analyses. To most observers, one
of them helps to confirm the other, although to some observers, two wrongs cannot be combined to make a right. Gene
Shoemaker has a different view. He now accepts that the
crater ages are clustered in time and is at least willing to
consider the idea that impacts are causing extinction. But he
suggests that this cause-and-effect link voids one's ability to

use the cratering and extinction analyses as independent lines of evidence.

ARE NEMESIS AND DINOSAURS INDEPENDENT?

If the Nemesis idea or any of the other astrophysical hypotheses pans out, does this add further proof to the original Alvarez proposition that the terminal Cretaceous extinctions were caused by impact? To what extent does each of the several theories I have been talking about depend on or augment the others? As you can imagine, things have gotten a bit confused with so many ideas flying so fast.

We can separate several elements of the problem. Since 1980, there have been five distinctly different propositions, as follows:

1. The impact of a large meteorite (comet or asteroid) 65 million years ago was the direct or indirect cause of the extinction of the dinosaurs and many other late-Cretaceous organisms.

2. Several other mass extinctions were also caused by large-body impact (late Eocene, Middle Jurassic, late Permian, late Devonian, and latest Precambrian).

3. Major extinction events over the past 250 million years are evenly spaced, every 26 to 30 million years.

4. Impact craters on earth are also periodic and approximately in phase with the extinctions.

5. Periodic extinction and cratering are driven by a clearly defined extraterrestrial phenomenon (companion star or Planet X or the changing position of the Sun in the Galaxy).

Any or all of these propositions may be dead wrong. We have seen that each has its detractors. Several competent scientists find more evidence for volcanism at the K-T boundary than for meteorite impact. Each of the other five impact-extinction cases has problems of fact or interpretation. The analyses of both kinds of periodicity have been strongly criticized. And all three astrophysical scenarios are debatable.

My own confidence in the five propositions varies, sometimes with the time of day. My hunch, however, is that their likelihood of being correct decreases in the order in which I have listed them, with impact at the K-T boundary as the cause of extinction being the most nearly proven.

If all five propositions are actually correct, we have an excitingly consistent general model for biological extinction: extinction caused by periodic comet showers driven by a regular solar-system or galactic motion. But in the more likely event that one or more of the propositions is wrong, what effect does this have on the others?

I see it this way. Impact as a cause of extinction can stand alone. We know that large objects have hit the earth throughout its history. There is no reason that they have to be regularly periodic. So if periodicity and Nemesis fall, the K-T impact is unscathed. By the same token, periodicity of extinction does not really depend on any of the other hypotheses, even though the other hypotheses are supportive. Periodicity could be real but driven, as Al Fischer suggested, by cycles of convection within the earth. While I don't find this idea very appealing, it cannot be ruled out absolutely.

One of the problems with the debates over the past couple of years is that people have tended to conflate the several propositions. The whole business is highly emotional, and the results of research on one of the problems tend to exert authority over the others. Suppose, for example, that Rich Muller finds Nemesis just where it should be and with the predicted orbit. I am sure that this would eliminate the opposition to impacts and periodicity almost overnight. And this would in some sense be legitimate, because the other elements would have been used successfully to predict a critical observation. But, I suggest, the reaction would be more emotional than rational.

Resolution of any of the problems I have been talking about will have a profound effect on the standing of the others—in spite of the fact that any claim of interdependence is not really valid.

Later press reactions, with increasing emphasis on debate and controversy. Most accounts stressed the uncertainties surrounding periodicity and the Nemesis-type theories based on periodicity.

10: *Role of the Press*

SAGANIZATION

RESEARCH scientists are ambivalent toward the press. Many colleagues complain bitterly about the coverage given to other scientists, opining that publicity degrades scholarship and is generally unsavory, and then brag of their own brief appearance on the evening news.

Past generations in science have been leery of public laundry, clean or dirty. My father told me recently that in the 1920s, he and many of his colleagues were reluctant to attend annual meetings of their own scientific societies. It was a little too public! An element in this was a cultural attitude among scientists, especially in academe, that the ideal was a sort of monkish community of dedicated scholars living in genteel poverty. Most of these attitudes disappeared after World War II, and especially after the launch of *Sputnik* in 1957 and through John Kennedy's great support of American science in the early 1960s. Now, the American scientific community tolerates, even encourages, the owning of BMW's (though probably not Cadillacs) and going to lots of meetings. Still, however, considerable suspicion is directed at the public side of science.

I think the American public is also ambivalent toward science. On the one hand, the Einsteins and the Salks are revered and pure science is seen as important in the maintenance and improvement of our way of life. Pest control using pheromones to confuse the sex lives of insects is labeled a clever and beneficial alternative to chemical pesticides, while Senator Proxmire confers the Golden Fleece Award to muddleheaded scientists who take government money to study the sexual behavior of animals. Nowhere are these two attitudes toward science better expressed than in the pages of the *National Enquirer:* both approaches often appear side by side.

A research scientist who allows his or her name to be displayed prominently in the press is taking a risk. If this same scientist appears on Johnny Carson or hosts a television series, disaster may await. I am talking about what I call "saganization."

Carl Sagan, a superb astronomer, has made his mark in several fields of science. Even as a student at Chicago, he had a seminal effect on the university community by spearheading discussion and research on the origin of life and related topics. He got people from many different disciplines talking to each other. He is now the David Duncan Professor of Physical Science at Cornell and Director of Cornell's Laboratory for Planetary Studies. He founded *Icarus,* a major journal in solar-system astronomy, and his entry in *Who's Who* is one of the longest I have seen, with a seemingly endless list of academic honors and awards. He is a member of the American Academy of Arts and Sciences (no mean feat for a physical scientist), although not the National Academy. He maintains an active research program, now focusing on the possibility of the existence of life elsewhere in the Solar System.

Pick a biologist or geologist at random and ask: "What do you think of Carl Sagan?" The answers will not be uniform but they will contain a disturbing number of negatives. You will hear that Sagan is more interested in personal glory than science. You will learn that his "Cosmos" series was a disgrace because it showed too many shots of Sagan and because it had a strong religious overtone. He sells T-shirts. He spends all his time on the lecture circuit and never does any science. His biology is terrible. He isn't much of an astronomer—even though the speaker knows neither astronomy nor astronomers. And so on. This is saganization.

As far as I can tell from my own observations, none of the negative charges can be sustained. I happen to think the "Cosmos" series was excellent education that did much to promote public understanding of science. A Sagan–Phil Donahue interview once turned into a superb lecture in chemistry. The scientific talks I have heard Sagan give are good, state-of-the-art science. He attends scientific meetings regularly, takes part in the discussions.

A surprising number of scientists have been saganized. Stephen Jay Gould is another example. A fine and imaginative scholar, Gould is breaking new ground in evolutionary biology and paleontology. He also happens to write and speak very well, talents that enhance an interest in communicating with the general public. His saganization bothers him deeply, because his most important objectives are to contribute to his science and to be respected for it by his peers. Carl Sagan, Steve Gould, and other examples have led many in the scientific community to be very cautious of the press.

THE PRESS: SCIENTIFIC AND PUBLIC

I have already referred to commentaries and editorials published in *Science* and *Nature* on the Nemesis question. *Science* and *Nature* are among a small but influential group of scientific journals that publish a great deal about the business of science: its politics, financial affairs, and substance. Both weeklies have very large international circulations. One or both are read by a significant fraction of the scientific community, although percentages vary greatly from one discipline to another. Both have regular sections on opinion, commentary, and research news. The issue of *Science* on my desk now, for example, has an editorial on the use of pheromones, a review of the epidemiology of AIDS, a report on the finding of the *Titanic,* and a review of new theories on Saturn's rings. The current issue of *Nature* has discussions of industrial research in Japan, AIDS in Poland, education in Illinois, and Soviet libraries, and reviews of current research on muscle contraction and on the crystallography of the cold virus. These journals also carry a large number of original scientific papers.

The research reviews and commentaries are written by highly trained staff—most hold PhD. degrees in a science. *Nature* invites the practitioners themselves to write the summary pieces. This means that the quality of reporting is generally high. The fact that the staff writers are well-trained means, however, that they have a tendency to develop their own research opinions about the relevant scientific questions. This is rarely found in the popular press.

The scientific press is having an increasingly direct influence on the pace of science and perhaps on its substance. Everything has speeded up in recent years, because a staff reporter can attend a scientific meeting, interview the participants, talk to people elsewhere by phone, and have a story published in two or three weeks. And the staff reporters have to be almost superhuman to jump from one arcane subdiscipline to another so quickly. Against the backdrop of the extremely slow pace of ordinary research publication, the news reports have an important impact. They provide a common way for all of us to keep up on a volume of source literature much too vast to be read or even skimmed.

At the other end of the spectrum, there is the daily news-paper and television news program, and the weekly news magazine. All of them report science, but the coverage is uneven and the journalists are rarely trained in science, although the general quality has increased dramatically in recent years. There are some superb science reporters in the public press, including Walter Sullivan of the *New York Times* (considered the dean of American science journalists), George Alexander of the *Los Angeles Times,* Boyce Rensberger of the *Washington Post,* and a few others. All too often, however, major science stories are assigned to general-feature editors or whomever happens to be available. Understandably, the popular press stresses subjects with obvious human interest.

Curiously, science rarely has a regular, assigned section in the public press. Astrology, sports, gardening, and chess get much more consistent coverage. The *New York Times* has a rather loosely defined science section on Tuesday, but one has to hunt for science news in Sunday's Week in Review. My own feeling is that science means more than this to the reading public and deserves more systematic treatment.

Between the daily newspaper and the scientific journal lie the science magazine and the television documentary. *Scientific American* has been a standard fixture in doctors' waiting rooms for years, but this is fairly heavy going even for professional scientists reading outside their fields. In the past few years, however, quite a number of science magazines aimed at the general public have appeared and are apparently thriving: *Discover, Science Digest, Science 85* (or *86* . . .), *Omni,* and so on. Even *Popular Science* is doing more with science. These magazines tend to emphasize hot topics and notable personalities. With cover stories or "centerfold" treatment, they can saganize a very sober scientist overnight. Also, as must be true throughout journalism, the science magazines feed on each other's stories, so that the same thing appears over and over again. But the quality has been generally excellent and the public understanding of science has been enhanced.

Television documentaries on science are significant, although I think there are some special problems with this medium. There must be pictures. Pictures of objects or animals are fine—like snakes and comets—but television runs into trouble when it tries to photograph the scientists who are doing the research on the snakes or comets. How do you photograph someone thinking? Or arguing with a colleague? What do you do with a scientist who has no instruments with flashing lights or experiments that make loud noises? In my own experience over the past couple of years, the television people solve this problem by forcing as many scientists as possible to be photographed in a laboratory or in the field, preferably with hard hat and safety glasses. This distorts large segments of the natural sciences and has a tendency to bias the public support of science toward those fields or ways of doing science that have lots of gadgets or are done in dense jungles or on high mountains.

PRESS REACTIONS TO THE NEMESIS AFFAIR

By the time the Nemesis theory came along in late 1983, journalists were already familiar with the idea of catastrophic extinction of the dinosaurs by comet or asteroid impact. Dinosaurs have a perennial appeal, of course, and the catastrophic aspects of the Alvarez theory made them all the more appealing. With the Death Star, the press really took off.

I will hit here only a few high spots in order to give the flavor. Following George Alexander's excellent story in the *Los Angeles Times* (September 4, 1983), stories appeared elsewhere with increasing frequency. John Noble Wilford of the *New York Times* wrote up periodic extinction and Nemesis on December 11, and over the next few months a large fraction of the major newspapers and news magazines in the world covered the story. Included were the principal American dailies plus *Newsweek, Maclean's* (Canada), *Recherche* (France), *Die Zeit* and *Frankfurter Rundschau* (Germany), the *Canberra Times* (Australia), *India Today, China Daily,* and *Panorama* (Italy). There was even a full article in the *Economist,* and the news finally reached *Reader's Digest* with a reprint of a piece Rich Muller had written for the *New York Times* Magazine. Strangely, *Time* did not touch the story at all until its cover story of May 6, 1985.

There was even a delightful syndicated column by Ellen Goodman under the title "Musings of a Dinosaur Groupie." Although very much out of her normal field, Goodman's essay is both amusing and penetrating. It is so good, in fact, that it deserves more than a mention here. She said (in part):

... in the past year, two scientists at the University of Chicago reported that . . . disasters have occurred like cosmic clockwork every 26 million years over the past 250 million years, wiping out huge numbers of life forms. The dinosaurs were just the biggest, most memorable of the victims.

Now when I look at the evolution of these theories, I wonder whether every era gets the dinosaur story it deserves.

. . . scientists are . . . part of their culture, their times. At one moment or another they are open to a certain line of questioning, a path of inquiry that would have been unlikely earlier on.

The scientists of the nineteenth century—a time full of belief in progress—saw evolution as part of the planet's plan of self-improvement. The rugged individualists of that century blamed the victims for their own failure. Those who lived in a competitive economy valued the "natural" competition of species. The best man won.

The latest theories may reflect our own contemporary world view. Surely we are now more sensitive to cosmic catastrophe, to accident. Surely we are more conscious of the shared fate of the whole species. . . .

In that sense, the latest dinosaur theory fits us uncomfortably well. "Our" dinosaurs died together in some meteoric winter, the victims of a global catastrophe. As humans, we fear a similar shared fate.[1]

The Goodman column raises some extremely important questions about how science works, questions I will discuss at length in the chapter on belief systems in science.

There were also some amusing offshoots of the main story, such as the time a photograph of Jack Sepkoski and me appeared on the "Page 10" column of the Chicago *Sun-Times.* We shared the column with Richard Pryor, John Hinckley, Jr., Tennessee Williams' brother, and a picture of an unidentified hockey fan in the act of baring her breast to distract the Edmondton Oilers in a game against her Chicago Black Hawks. Another time Ian Warden of the *Canberra Times* described me as "part Henry Kissinger part Ronnie Corbett."

But how did the public press in general approach the story? There were a few standard elements. Dinosaurs were emphasized, of course, usually with drawings of common dinosaur types and often cartoons of dinosaurs being hit on the head by large rocks falling out of the sky. There were some good diagrams showing the Earth in relation to the Oort Cloud and the orbit of Nemesis. And there were some depictions of the geologic time scale, with the major extinction events indicated with sketches of animals that went extinct. This caused a bit of difficulty, however, because except for the dinosaurs, most of the extinct organisms are unfamiliar to the general public.

Most stories made the tie to the earlier work of the Berkeley group on Cretaceous extinctions, and as a result iridium and other indicators of large-body impact were

Page **10**

By Marla Paul, Ray Hanania, Lynn Sweet and Robert Feder

The end is coming—but not to worry

The discovery by University of Chicago paleontologists **David Raup and J. John Sepkoski Jr.** at first sounds like a vaudeville joke: The bad news is: The end is coming. The good news is: It's 13 million years away.

The duo from Hyde Park has been thrust from the obscurity of scientific circles into the limelight via recent issues of Time and the New York Times magazine. That's because of their role in a new explanation of why dinosaurs are extinct—and why human beings ultimately may suffer the same fate.

"We always thought of the dinosaur as dumb and deserving to be extinct," said Raup. Not true. Dinosaurs, it turns out, were merely victims of circumstances. As Raup, 52, and Sepkoski, 36, were surprised to find, certain life-forms are subject to mass extinction, and this happens roughly every 26 million years. A new life-form then arises. This caused quite a stir among other scientists to figure out why.

The new theory: Periodically, a curtain of comets covers the sky for thousands of years, preventing sunlight from reaching Earth for extended periods. Life is wiped out.

But for Sepkoski, who gardens and makes furniture between research and teaching, short-term problems are of greater concern. He cautions, "I think we have to be more concerned with the immediate dangers from our own perversions, nuclear war, overindustrialization and overpopulation. . . ." Raup figures that new technology eventually will divert any killer comets heading our way. And when he sails out of Jackson Park harbor on summer evenings, he says, his attention is on his wine and cheese, not on the stars. Notes the scientist, "It's more interesting."

For the SUN-TIMES—Frank McMahon

Paleontologists David Raup (left) and J. John Sepkoski Jr. know the end is coming—in 13 million years.

Jack Sepkoski and I make the "Page 10" column of the Murdoch-owned *Chicago Sun-Times.* Note the standard inclusion of our ages, neighborhood, and hobbies. Though the reporting is reasonably accurate, the inevitable glitches are there, such as saying that it is the comets themselves that block the sunlight in a mass extinction. We shared this particular column with Richard Pryor, John Hinckley, Jr., Tennessee Williams' brother, and a bare-breasted hockey fan. (© News Group Chicago, Inc., 1985. Reprinted with permission of the Chicago *Sun-Times*) covered. Virtually all journalists tried to check the story with scientists other than Jack and me and the Berkeley people. This almost always meant one or two quotations from scientists who were skeptical. For example, John Noble Wilford's first piece on the subject quoted Norman Newell, a paleontologist at the American Museum of Natural History, as saying that periodic extinction could be "a statistical artifact that shows order or pattern where none exists in reality." This opinion should have carried a lot of weight, because Newell, one of our most respected paleontologists, has done some of the best work on mass extinction. But such quotes typically were tagged onto otherwise upbeat articles. I doubt that they made much impression on the reading public. In other words, the public press supported periodicity in extinction and its extraterrestrial interpretations.

Several major science magazines published more complete treatments of the Nemesis story. *Discover* devoted a major part of its May, 1984, issue to a three-part series of articles. One went through the extinction periodicity in considerable detail, one described and analyzed the several astrophysical interpretations, and the third discussed the intriguing topic of possible measures to defend the Earth against incoming comets and asteroids. This segment ended with Rich Muller saying:

> Even if the next comet shower were to come in a couple of years, we have the technology now to handle it. We're the first species that has that capability.

In January of 1985, *Science Digest* ran a long section on what it considered the 1984 "Stories of the year." The claim was made that "The astronomy story of the year, perhaps of the decade, was produced by geologists and paleontologists." I can't help but feel a bit embarrassed by this, because 1984 was a good year for astronomy without any help from geologists and paleontologists.

THE TIDE TURNS

As 1985 progressed, the tone of the reports took on quite a different character. The new theories slipped into the background and controversy moved to the fore. By this time, periodicity and the several astrophysical explanations had been scrutinized by the scientific community and counterarguments were being developed. Many people did not like the Raup and Sepkoski statistical analysis. Also, the two major extraterrestrial interpretations, the companion star and our position in the Galaxy, were coming under fire. The alleged periodicity in the ages of impact craters was being debated heavily. The quoted statements by prominent paleontologists, geologists, and astronomers became less cautious and more strident.

One especially significant sequence of events began early in 1985. In January, Scott Tremaine of MIT participated in a special session devoted to Nemesis and related problems at a major meeting of astronomers in Tucson. Among other things, he reported work he had done with a graduate student, Julie Heisler, indicating that periodic extinction was not on solid ground statistically. Richard Kerr of *Science* then described Tremaine's work in a column entitled "Periodic Extinctions and Impacts Challenged." Kerr's column had quite an impact itself. Many readers jumped to the conclusion that the K-T impact of the Alvarez group was being challenged as well as periodicity. Dick Kerr had no such message: his title had contained an unfortunate ambiguity.

More important, the Kerr piece almost certainly led to the *New York Times'* first editorial on the subject (April 2, 1985). The editorial ended with the following:

On closer scrutiny, the alleged repeating pattern of mass extinctions has faded. The dinosaurs and other vanished species didn't all turn feet up in a day; some were in decline before the end of the Cretaceous. The thin layer of iridium that has been found in many geological strata dating from 65 million years ago could indeed have come from a meteorite, as the Alvarezes suggest, but eruptions of volcanos are now known to be sources of iridium too.

Terrestrial events, like volcanic activity or changes in climate or sea level, are the most immediate possible causes of mass extinctions. Astronomers should leave to astrologers the task of seeking the cause of earthly events in the stars.

THE NEW YORK TIMES, TUESDAY, APRIL 2, 1985

Miscasting the Dinosaur's Horoscope

During the close of the Cretaceous era some 65 million years ago, all dinosaurs disappeared from the earth. Paleontologists, the students of fossil life forms, have for decades debated inconclusively the reasons for that extinction, but five years ago their game was suddenly snatched away by two brash Berkeley scientists and a crowd of astronomers.

Luis Alvarez, a physicist, and his son Walter, a geologist, contended that a meteorite had slammed into Earth raising such a storm of dust that the sun was blotted out and whole species of animals fell extinct worldwide. Stretching a provocative idea even further, other scientists claimed to discern a regular pattern in the fossil record: mass extinctions every 26 million years.

The notion of regular extinctions got astronomers excited because the deus ex machina required to make giant meteorites crash into earth like clockwork every 26 million years clearly lay in their province. Some posit that an unseen companion of the Sun, christened Nemesis, shakes loose comets each time its orbit passes near a comet cloud. Others contend that the Sun, as it bobs up and down through the plane of the galaxy, is buffeted by comets or dust clouds.

These are rich hypotheses. Why, then, without any further evidence, do they seem so unsatisfying? Perhaps because complex events seldom have simple explanations. Invoking regular squads of meteorites to dispose of the dinosaurs and other vanished species is only to exchange one mystery for another.

On closer scrutiny, the alleged repeating pattern of mass extinctions has faded. The dinosaurs and other vanished species didn't all turn feet up in a day; some were in decline before the end of the Cretaceous. The thin layer of iridium that has been found in many geological strata dating from 65 million years ago could indeed have come from a meteorite, as the Alvarezes suggest, but eruptions of volcanos are now known to be sources of iridium too.

Terrestrial events, like volcanic activity or changes in climate or sea level, are the most immediate possible causes of mass extinctions. Astronomers should leave to astrologers the task of seeking the cause of earthly events in the stars.

The first *New York Times* editorial attempting to discredit extinction by impact, periodicity in extinction, and Nemesis. The editorial ends with the now-famous claim that this research should be left to astrologers. To many observers, the editorial would have been better suited to April 1st than April 2nd. (Copyright © 1985 by the New York Times Company. Reprinted by permission.)

These are such remarkable statements that a number of my colleagues suggested that the editorial had been scheduled for April 1st but missed by a day.

You will notice that the *Times* editorial attacks the K-T impact itself as well as periodicity, and in so doing dredges up a number of standard arguments, including the volcanic alternative and the fact that dinosaurs had been in decline for some time before the end. Much more surprising, shocking in fact, is the final sentence about the foolishness of astronomers who try to find causes of "earthly events in the stars." It says, in effect, that everyone knows that the Earth could not be influenced by its cosmic environment and only astrologers would be so naive as to make the attempt. This, unfortunately, reflects the stance of many of my more Lyellian colleagues. I am glad to report that Walter Sullivan had nothing to do with the editorial.

The *Times* editorial itself probably had little direct effect on opinion in the scientific community. Its anti-intellectual tone was all too obvious. But it did stir up the popular and scientific press.

Jack Sepkoski and I were deluged with questions about Scott Tremaine's analysis of our data, which had led indirectly to the *Times* editorial. Would we respond? Could we respond to an MIT physicist? This got a bit messy. The Tremaine analysis was technically hearsay, because it did not exist in the conventional sense of a scientific publication. To be sure, he sent us a copy of the manuscript shortly before submitting it for publication in a special volume based on the Tucson meeting. We were working on a response but could not say anything substantive about it publicly, for fear of having our own paper on the subject disqualified by prior publication in the press. Besides, we had nothing to rebut until Tremaine's paper had been reviewed, revised, and finally published. At the end of 1985, a year after Tremaine's statistical attack, his paper had yet to be published.

Tremaine's analysis was basically excellent. He had found a problem that we knew of and had tried to treat in our *PNAS* paper. But Tremaine's approach, while much more elegant, unfortunately contained a significant mathematical error that flowed from his lack of familiarity with paleontological data. As I write this, Jack and I have a manuscript on the problem in press with *Science* and I should not say more about it here.

TONI HOFFMAN

In the late spring of 1985, a bombshell hit. Antoni Hoffman, a Polish paleontologist working at Columbia's Lamont Geological Observatory, published a densely packed critique of extinction periodicity in *Nature*. John Maddox, *Nature*'s chief editor, preceded the Hoffman paper with a laudatory News and Comment piece. Maddox said, in effect, that the recent love affair with catastrophism was dead.

Hoffman's criticisms, which had been kicking around for some time, had already been largely published elsewhere. Even so, the force of publication in *Nature* plus John Maddox's endorsement had an enormous effect on the scientific community. I suspect that few people actually read Hoffman's paper. The Maddox commentary was enough for busy people in other fields of science. If I had been looking in from the outside, my reaction would have been the same. What to do? Again, the problem was trying to avoid responding in the press. Both Jack and I had a series of very difficult conversations with people in the scientific press who felt obliged to write up the "event." They wanted our reactions. We were on a tightrope trying to convince the reporters that there was much more to the story but without giving any of the substance of our arguments. With friends, we were not nearly so inhibited!

Roger Lewin of *Science* wrote a superb balancing act of

his own under the title "Catastrophism Not Yet Dead." Based largely on his own scientific knowledge, Roger presented a few counter-arguments to Hoffman and generally suggested that it was too early to write the obituaries on periodic extinction and Nemesis. He ended with: "The new catastrophism may well have to be abandoned, but not yet."

The Hoffman article led to yet another *New York Times* editorial, this one titled "Nemesis of Nemesis." It was a more sober piece, mostly devoted to Toni Hoffman's thesis that Nemesis and periodicity were done for. As the *Times* put it, "The basis of these theories may have crumbled into statistical dust." The editorial concluded with another crack at the original Alvarez hypothesis:

> If that analysis holds up, all candidates for periodic extinctions expire. A chance meteorite may have polished off the dinosaurs. But until its place of impact is discovered, it's just as well not to rule out terrestrial suspects—like change of climate.

Attaching significance to the missing crater is a cheap shot. Because of the high probability that a 65-million-year-old crater would be lost by subduction of the ocean floor or by normal erosion, the lack of a crater is of minor importance.

Steve Gould came to our rescue with an essay in *Discover* (October, 1985) under the title "All the News That's Fit to Print and Some Opinions That Aren't." He showed very elegantly that Toni Hoffman did not have a convincing case. And then Steve went after both of the *New York Times* editorials and ended with two delightful parodies of the April 2, 1985, editorial. One of them purported to be taken from an editorial in the *Osservatore Romano* of June 22, 1663, and read as follows:

> Now that Signor Galileo, albeit under slight inducement, has renounced his heretical belief in the earth's motion, perhaps

students of physics will return to the practical problems of arma-
ments and navigation, and leave the solution of cosmological
problems to those learned in the infallible sacred texts.

In the months since the second *Times* editorial, the pace
of charge and countercharge has become almost frantic.
Hardly a week goes by without a new paper in either *Science*
or *Nature* reporting a new argument or line of reasoning.
The credibility of the original galactic-plane idea has been
analyzed by astrophysicists from several new viewpoints.
Several new studies appeared detailing extinction in specific
parts of the fossil record. Rich Muller continues to search the
sky for Nemesis. The soot story of the Anders group will
probably start up interest anew—especially with its implications for nuclear winter. The press is reacting to many of the
new developments as they come, hoping, I think, finally to
gain a confident answer.

THE PRESS: GOOD OR BAD INFLUENCE?

When all is said and done, it is hard to say whether the press,
public or scientific, helps the progress of science and whether
it does a good job of informing the general public.

On the one hand, progress in research has been enhanced
by the quick transfer of information between scientific disciplines that good reporting has accomplished. And a lot of
time has been saved. The best example of this in the Nemesis
Affair is the initial reporting of Jack's Flagstaff presentation
in late 1983 by *Science, Science News,* and the *Los Angeles
Times.* For busy scientists, the news report or commentary
in a scientific journal may be the only way to learn what is
going on in other fields.

On the other hand, the press promotes a lot of misconcep-
tions and spawns numerous mistakes. We have seen a number of examples. An editor or staff writer has to make quick

decisions about new developments in science—without the benefit of thorough testing and evaluation. On the whole, I think they have done a remarkably good job, but mistakes have been and probably will continue to be made. On balance, the positive results probably outweigh the mistakes. And it is surely good for the ego when one's paper is introduced by an editorial commending the work. This has happened to me often enough to let a few nasty editorials go by.

If the press has one systemic problem, it is the desire to present the definitive answer in each article. To be sure, there is generally new information and, in an ideal world, this new information sets the record straight: the conclusions will hold indefinitely. And most news articles are written from this viewpoint. But in the case of Nemesis, as with so many situations in and out of science, new information is ephemeral and often wrong. On a controversial issue, the wind shifts repeatedly. Through journalistic desire to present the final word, the reader is pushed back and forth between article and article. Confusion reigns. Journalism may well move at a pace far beyond the capability of scientific research to keep up. It will be helpful when the press recognizes this problem and figures out some way to report new developments while eschewing final answers.

Many have argued that newspapers have no business writing editorials on purely scientific questions. Some say that the editorial writers should leave the business of science to the scientists. I disagree. Scientific issues are of both public interest and societal importance. To be sure, most editorial boards of major newspapers do not contain card-carrying scientists, but they also lack expertise in political science, epidemiology, and most other technical fields. If we deny them opinions about scientific matters, we must also rule out AIDS and theories of taxation.

11: *Onward to the Earth's Magnetic Field*

WHAT follows is a short shaggy-dog story—an offshoot of the Nemesis Affair that may or may not have significance in the long run. It is still too early to say. But the story is a useful vehicle to describe some interesting aspects of the research process and contains some oddities and a few surprises. Also, it is an aspect of the Nemesis Affair that consumed much of my energies for about a year.

MAGNETIC REVERSALS

For most of us, the Earth's magnetic field is little more than a convenience in navigation. We don't need the magnetic field very often. On any reasonably sunny day, the position of the Sun provides an intuitive sense of north and south, and even this is unnecessary when we are in familiar surroundings. Having no biological ability to sense a magnetic field, we do not use it without mechanical aids. But to many organisms, the ability to sense magnetic fields is perhaps as important as the other senses. Quite a variety of fish, birds, and insects actually manufacture tiny magnets (as part of their metabolism) and use these in conjunction with good "onboard computers" to detect subtle aspects of the magnetic field. They are able to sense a magnetic topography every bit as real as the solid topography so comfortable to human beings.

In another arena, one of the triumphs of modern geophysics was the discovery that the Earth's magnetic field occasionally reverses. That is, it flips: north becomes south and south becomes north. This discovery came about because certain rocks retain a record of the magnetic field in existence at the time they were formed. From this, a geologic chronology of changes in the direction and intensity of the Earth's field has been constructed.

Shortly after magnetic reversals were recognized (in the late 1950s), the question naturally arose: What would the environment be like during a magnetic reversal? Would biological consequences flow from the intensity of the field's drop to zero during the reversal? Could this produce a mass extinction? These questions were followed by some rather abortive research. When only a few magnetic-field reversals had been found, there were cases of what appeared to be a close association between them and times of extinction. But these disappeared as more reversals turned up.

There was no compelling reason to expect reversals to cause significant biological effects. Navigational use of magnetic fields is probably not important to most species. On the other hand, if the Earth has no magnetic field for a time, the shielding effect of the Van Allen belts will be lost and organisms will be subjected to slightly higher levels of cosmic radiation, so that rates of gene mutation may increase. But it was soon shown that this effect would be biologically trivial and the whole matter was more or less dropped.

In spite of the lack of promise, the idea of a connection between reversals and extinction has been kicking around in the backs of many people's minds. We know appallingly little about the physiological effects of magnetic fields. Very few experiments have been done with laboratory animals living in the absence of a magnetic field. Furthermore, even though geophysicists have done a splendid job of describing and documenting the history of magnetic reversals through geologic time, little is known for sure about how or why the field flips from north to south. Could a large-body impact jolt the Earth sufficiently to reverse a delicately stable magnetic field? This has been suggested in the geophysical literature a couple of times but nothing definitive has come of it.

A RESEARCH EXCURSION

By the summer of 1984, it had become clear that our analysis of periodic extinction was fundamentally limited by the powers of statistical inference. We trusted our results, but we needed independent corroboration. It was important to find out whether the extinction argument had predictive value in the sense of being associated with other signals in Earth history that might be reacting to the same stresses. Many people were looking at impact craters and analyzing rocks for additional chemical signals. We set out to find other signatures of promise in the record. I decided to look at the magnetic-reversal record. Was it also periodic, and if so did the periodicity match that of extinction?

I had no business getting into geophysics (in spite of being chairman of a department called Geophysical Sciences). I have only anecdotal knowledge of magnetic fields. My risk of blunder was obvious. I decided to gamble. As you will see, it is not yet clear whether the decision was a wise one.

Unlike the paleontological record, the reversal record is fairly clean, at least for the past 165 million years of geologic time. About 300 complete reversals have been found in this interval, most well dated. There are some very long-term trends in the number of reversals per million years, and these have been known and accepted for many years—even though there are no good explanations for them. My interest, however, was in the fine structure of the temporal pattern: minor variations superimposed on the broader changes.

I was encouraged to find a recent paper claiming periodicity in the number of reversals through time. This was a study by two Indian geophysicists, J. G. Negi and R. K. Tiwari, which concluded that there is a 32-million-year periodicity in the magnetic data. Their methods did not allow one to say whether the pulses of reversal activity came at the same time as extinctions. Later, as my work progressed, I found another paper: a French group, headed by A. Mazaud, claimed a 15-million-year periodicity positioned in time such that every other pulse was a bit stronger and fell approximately at one of our extinctions. Very tempting!

On the other hand, the problem had been looked at by a number of other geophysicists over the years. All agreed that the fine structure of the reversal record showed only random fluctuations in the number of reversals per time interval. They had been unable to reject a hypothesis of randomness. While this does not prove randomness, it puts the burden of proof on one who says the pattern is non-random. In fact, randomness was the strong conventional wisdom in geophysics—and may still be. But conventional wisdom has a way of being wrong and the Indian and French results gave some hope.

To make a long story short, I did some elaborate statistical analyses of the magnetic record, using basically the same techniques Jack and I had used for the extinction data. I found an impressive 30-million-year periodicity that matched the extinction periodicity fairly well. I say "fairly well" because no 30-million-year cycle can stay in sync with a 26-million-year cycle for very long. The most recent extinction events are at 11, 38, and 65 million years B.P., and their counterparts in the magnetic record are at 10, 40, and 70 million. There is enough uncertainty in the statistical analysis that the difference may not mean much. Remember that

Rampino and Stothers at NASA got a 30-million periodicity from Jack's and my extinction data, and the Berkeley group found a 28-million-year periodicity in crater ages. All of these studies could be saying the same thing.

I hope you will appreciate the possibilities here for the research to be "led" by expectations. I have already mentioned the cynical saying: "I would not have seen it if I had not known it was there." The dangers of unconscious bias are always present. No statistical test procedure protects science against preconceptions. This is by far the most daunting part of research.

PEER REVIEW

The statistical results were firm enough, and the possible connections to biological extinction so intriguing, that I wrote up the work as a short paper for *Nature*. Peer review in this case would be a good test, although much depended on the editor's selection of reviewers.

The paper was submitted to *Nature* in September of 1984. At the same time, I sent copies to a dozen people for their feedback. One could argue that I should have gotten this private feedback before submitting the paper—but I was impatient. The initial review process was completed by *Nature* in mid-November. I was informed that the reviews were conflicting: one recommended publication and the other urged rejection. When this happens, the editor of the journal has a problem and several options. Often, the manuscript will be sent, with the reviews, to a third reviewer in hope of resolving the issue. Or the editor may make a yes-or-no decision himself.

In this case, *Nature* chose a third alternative: it neither accepted nor rejected the paper but returned it to me (with the reviews) to revise and resubmit. At the same time, I was invited to suggest the names of six additional reviewers.

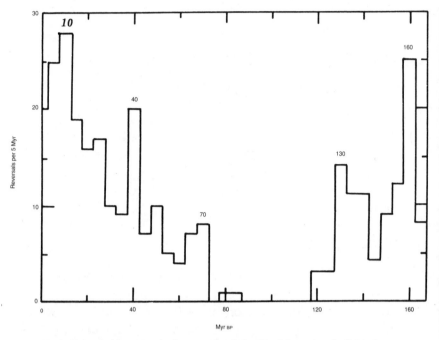

Periodicity in the record of reversals of the Earth's magnetic field, from my 1985 *Nature* paper. The histogram shows the number of complete reversals in five-million-year (myr) intervals for the last 165 million years. The general trend (from right to left) shows a decrease in the reversal rate, culminating in an interval from about 120 myr BP to about 80 myr BP with no reversals (the Cretaceous quiet zone), and a resumption of reversal activity with an increase to the present day (left). My *Nature* paper argued that there is a regular 30-million-year periodicity superimposed on the larger trends, with pulses of activity centered on 10, 40, 70, 130, and 160 myr BP. (After D.M. Raup, 1985, *Nature, 314:* 341–343, fig. 1)

It is not uncommon for editors to ask authors to suggest the names of several people to whom a manuscript might be sent for review. This is an interesting aspect of peer review. The process is generally secret and the author of a paper usually never knows who reviewed the work. But editors find the suggestions of the authors useful and they often choose one or more from the list provided. Naturally, the author has an interesting time drawing up the list. It can become a nerve-racking crapshoot if taken too seriously.

In this case, I knew the identity of both reviewers. One, a geophysicist in Australia, had signed his review and the editor had chosen not to cut off the name. The other is a prominent geophysicist in California to whom I had already sent a copy of the paper. He replied saying that he had received the paper also from *Nature*. The California reviewer was one of those who had looked at the problem some years before and had concluded that the small variations in the magnetic record were random. His review was very positive and recommended publication. He made clear, however, that he was uncomfortable with my results even though he could find nothing wrong with the methods.

The Australian reviewer found the paper totally unacceptable, saying it contained a "serious fallacy" and that "the conclusions drawn in the paper are unjustified." The four-page review was extremely hard-hitting. As it turned out, this reviewer's criticisms were very helpful. I used them to make major revisions of technique and presentation. The peer-review process was working as it should but so often does not.

Having revised the manuscript, I sent it back to *Nature* along with the requested six names of suggested reviewers. In my cover letter, I urged the editor not to try the Australian again, quoting some passages from the negative review to show that the Australian's mindset destroyed his objectivity. *Nature* promptly sent the revised manuscript back to the Australian reviewer! In addition, they sent it to one of the people on my list: Timothy Lutz at the University of Pennsylvania. Although I did not know Lutz personally, he was a very bright young geophysicist with experience in time-series analysis.

In late January of 1985, I was informed by *Nature* that both reviewers had approved the revised paper and that it was being returned for minor, technical revision before going to press. I was, of course, delighted. The peer-review system had worked to improve the research and the whole process had taken less than five months. Little did I know what was coming.

A month later, over a delightful lunch in Canberra, the Australian reviewer said he was still doubtful of my results but could find nothing wrong with the analysis. We parted with a five-dollar bet on the ultimate outcome—a bet that has not yet been settled.

LUTZ REBUTS

The 30-million-year-periodicity paper was published in late March. I waited for reactions. The press gave the paper almost no attention but enough in the scientific community were by now sensitized to reading about geological periodicity that I was sure the right people were reading the paper. Tim Lutz had gotten interested in the problem and was doing additional analyses of his own. We met for the first time, at a symposium at Princeton honoring Al Fischer, and worked with strategies for further study of the magnetic record.

In his original review for *Nature,* Tim had suggested some different approaches and I had considered holding back, with the idea that the two of us might collaborate on a more complete treatment. After I decided to go ahead with publication alone, Tim continued his own analysis.

In late spring, Tim Lutz completed the extended analysis and submitted a paper to *Nature* under the title "A Reappraisal of Periodicity in Magnetic Reversal Record." He forwarded a copy to me at the same time. Not surprisingly, *Nature* also sent it to me for review.

The Lutz paper was an elegant piece of scholarship and showed rather effectively that although the 30-million-year periodicity might exist in the magnetic record, my paper had not proven it. What to do? First, I applied his general approach to my version of the data with my own computer programs. The results were the same as his. I had no choice but to recommend the Lutz manuscript for publication.

This raised a problem vis-à-vis periodic extinction. My methods for magnetic reversals were basically the same as those Jack and I had used for extinction. If one study was wrong (or at least inconclusive), was the other also wrong? This could be very serious, because of the highly controversial nature of the extinction business. Remember that the whole periodicity-Nemesis question was being debated furiously at this time. Lutz had noted the problem in his paper but suggested that the extinction data were sufficiently different that his criticisms might not apply. Fortunately, he was right: applying his new approach to the extinction left the periodicity unscathed.

RETRACTION

At this point, I had several options, as did the editors of *Nature.* The Lutz paper could be published as a straightforward rebuttal to my paper, in which case I would have right of response. Or Lutz's paper could be published as a freestanding article, in which case any reply from me would have to come in a later issue, with Lutz having right of response. In either scenario, I would have two objectives: to acknowledge the effectiveness of the Lutz analysis and to protect the extinction analysis from being declared guilty by association.

Actually, the matter turned out differently. A couple of good conversations with Peter Gambles, the editor in *Nature*'s London office who had been handling the case, led to the following solution: Lutz's manuscript to be published as a free-standing article with a News and Views piece by me introducing it in the same issue. This was Dr. Gambles' suggestion and I was grateful for it. It gave me the opportunity to acknowledge the Lutz work while at the same time "protecting" periodic extinction. News and Views is the same prominent spot in *Nature* that John Maddox and others had used to comment on Toni Hoffman's paper and other aspects of the Nemesis Affair.

As I write this, the issue of *Nature* with the two pieces has just been published, so one can only guess how the scientific community will react. It is not at all clear whether the combination of Lutz's re-analysis of the magnetic record and my retraction will help or hurt periodic extinction and Nemesis. But the plot actually became thicker in the months during which all this writing and reviewing was being done.

After sending Peter Gambles my News and Views text, three things happened:

1. I received yet another manuscript from *Nature* for review. Written by two respected British geophysicists specializing in the magnetic-reversal record, it reported a 30-million-periodicity in that record. The startling thing was that the authors had somehow missed my paper and were evidently quite unaware of what I had published a few months before in the same journal. This, added to the original Indian and French studies, gave me renewed confidence in my results. Was I too quick to retract? Unfortunately, the new paper had a technical problem, in that the statistical significance of the results had not been tested, and *Nature* rejected it. So, it ceased to exist and will remain nonexistent unless a revised version is submitted and published. (In fact, the

paper has been resubmitted but not yet accepted or rejected.)

2. Mike Rampino and Dick Stothers at NASA heard about my impending retraction and did their best to talk me out of it. They had done parallel work on the magnetic data and had come up with the same results but did not publish them, because I had done it first. Now, they had worked with Lutz's study and did not find it convincing. Rampino and Stothers think the magnetic pattern is real and will probably write a Matters Arising piece to rebut Lutz.

3. Rich Muller at Berkeley, one of the fathers of Nemesis, has a paper in press that builds on the magnetic-periodicity arguments. Rich and a colleague, Donald Morris, think they have a credible mechanism whereby comet or asteroid impacts could cause reversals of the Earth's magnetic field. Ironically, the mechanism calls for rapid sea-level change as an integral part of the process. Naturally, Rich hopes that the magnetic periodicity holds up. His abstract ends with the optimistic statement:

> The model can account for the previously mysterious correlation between geomagnetic reversals and mass extinctions.

OUTCOMES

At this point, I think you will agree that the question of magnetic periodicity is a mess. We don't know whether the fine structure of the magnetic record shows periodicity or not, and even if it does, we don't know whether the reversals are related to biological extinction or to large-body impact or both. And we don't know what effect this confusion will have on the credibility of the original work on periodic extinction and Nemesis.

Although this situation is confused, it is not at all atypical of scientific research, except perhaps for its rapid pace and the large number of people involved. The process of arriving at a truth, or at least a consensus, is often a highly complex and rather disorganized business of backing and filling. It is a far cry from the stereotyped kind of science where a simple experiment or measurement establishes a discovery once and for all and without argument.

Actually, there is a spectrum of situations between the very simple solution, quickly and neatly achieved, and the uncertain and difficult sort of case like magnetic periodicity. For example, many of the difficulties surrounding questions of genetic differences between races, the epidemiology of AIDS, and even Einstein's relativity are examples of the "difficult" end of the spectrum. The discovery of the genetic code may be an example of the "clean" or "easy" end, but the elegant and irrefutable solution to the DNA problem was preceded by many frustrating years of mistakes and blind alleys.

I should also emphasize that all the players in scientific research are people. I have left out a lot of the purely personal and emotional aspects of the magnetic-reversal story, but I can assure the reader that this side of science is important. It has probably been described best by James Watson in *Double Helix,* the story of his discovery (with Francis Crick) of the structure of DNA.

12: Belief Systems in Science

In this chapter, I want to build on the Nemesis experience to say something about the more subtle aspects of doing scientific research, setting forth my view that working scientists are subject to many more prejudices and preconceptions than is generally thought.

The structure of science, with its rigid procedures and standards for testing of hypotheses, helps to organize our knowledge (and ignorance), and the results of the "scientific method" are probably more objective and dispassionate than in most other fields of human inquiry. But the process still contains strong emotional and sociological elements. I will use a number of examples, including the Nemesis story, to explore this non-objective side of science.

I must say emphatically at the outset that I am not professionally competent in the history and philosophy of science. My remarks will try to describe and comment on science as I see it and will not presume to second-guess such scholars as Kuhn and Popper. One of my main objectives is to temper some popular stereotypes about the way science works on a day-to-day basis.

THE SCIENTIFIC METHOD

Almost all scientists think they know how science differs
from philosophy, religion, and plain guesswork. Or, at least,
all scientists I have talked to think they know. But it is quite
a different matter to find out just how science differs from
other forms of inquiry. To some, science is defined by the use
of experiments to test the predictions of hypotheses. To others, it is any kind of careful scholarship not burdened by a
prior commitment to a particular answer or belief system.
Most scientists argue that religion is not science because
religion involves no experiments, tests no hypotheses, and is
committed beforehand to a set of beliefs. Science and faith
are antithetical. It follows that scientific research is objective
because the scientist is not influenced by prior expectations
and is willing to let the chips fall where they may. I think
these statements contain a fair amount of bunk.

GUILTY UNTIL PROVEN INNOCENT

All sciences work within theoretical frameworks. These are
alternatively called models, paradigms, hypotheses, concepts, principles, or precepts. Whatever they are called, they
provide basic frameworks for thinking about problems and
for interpreting observations. The Darwinian theory of evolution is an example of such a model or paradigm. The
Nemesis theory is another example, albeit at a quite different
scale. Every field of science has one or more theories of this
sort.

In nearly all cases, the theoretical frameworks are
"thought up" by imaginative scientists. They usually start
out as little more than hunches—perhaps suggested by observational data but not required by them. The periodic table
of the elements was not forced on chemistry by an immense
weight of evidence that allowed no other interpretation.

Once a new theory or paradigm is proposed, it is tested and evaluated by the scientific community—assuming it is reasonably credible and evokes enough interest. New or old observations or experiments are evaluated in light of the new idea. If they fit, the new idea is supported. If they don't fit, the new idea gets a black mark. Rarely is there a single crucial experiment to prove or disprove the new idea.

All this is straightforward and has its counterparts throughout human affairs. But there is a tricky aspect. The new idea is almost never in a vacuum: one or more other ways of describing or explaining the same phenomena already exist and are accepted by the scientific community. Walk up to an astronomer and ask: How was the universe formed and how old is it? You will always get an answer, regardless of the decade or century in which you ask the question. Yet the answers change as new ways of interpreting the observations arrive and become established.

Acceptance of any new theoretical framework depends on credibility and improvement on existing alternatives. In cases where the choice is not immediately obvious, the burden of proof generally lies with the new idea. This is very important. Given a choice, the scientific community invariably sticks with the conventional wisdom. Furthermore, the older ideas have usually been around long enough to have accumulated supporting evidence, whereas the new idea rarely has much going for it, at least at first. It is not a fair game.

Despite the lack of "fairness" to new ideas, the traditional practice in science may serve us better than a more democratic mode. History shows that most new ideas fall. Science would be very confused much of the time if all new ideas were given precisely equal treatment.

It is clear, therefore, that the new theory is guilty until proven innocent, and the pre-existing theory is innocent until proven guilty. Consider the quote I used from one of Bill

Clemens's papers in Chapter 4. When considering whether the Cretaceous extinctions were caused by a meteorite impact, he said:

> . . . analyses of the paleobiological data suggest that such an event is not required to explain the biotic changes during the Cretaceous-Tertiary transition.

The operative word here is "required." Clemens was not saying which explanation of the extinctions is the most likely. Rather, he was saying that we already have acceptable explanations without resort to meteorite impact. The practical effect is that a new idea requires a truly compelling case—intellectual overkill—in order to displace the incumbent.

Although I am somewhat uncomfortable with the procedure I have just described, it is a strong element in the culture of science and may even move us ahead in the long run. But it has a sometimes crippling effect on progress, especially if the conventional wisdom in a field has been established for a long time with its "innocence" supported by a great variety of evidence. The persistence of the Ptolemaic system of astronomy, whereby the Earth was at the center of the universe, is a classic example. It was accepted by virtually all educated people, partly because of inertia and partly because it worked remarkably well as a predictor of eclipses and so forth. Its replacement by the Copernican system was a long and difficult process. By the same token, Einstein's theory of relativity had to withstand years of derision by his fellow physicists.

Perhaps the only thing that saves science from invalid conventional wisdom that becomes effectively permanent is the presence of mavericks in every generation—people who keep challenging convention and thinking up new ideas for the sheer hell of it or from an innate contrariness.

The Nemesis Affair shows many examples of the conflict

between new and old theories. But it is virtually impossible to prove the guilty-until-proven-innocent bias. When an article contains a list of ten reasons why the K-T extinctions could not be caused by impact, there is no way to know to what extent the author is consciously or unconsciously selecting or slanting the data. With regard to the original Alvarez proposal of the K-T impact, I am reasonably sure that protection of the conventional wisdom has played an enormous role. In the absence of this conventional wisdom, the impact (and its biological effects) would now be seen as established beyond any reasonable doubt.

WINNING A LOTTERY: SCIENCE OR RELIGION?

A curious incident exemplifies the great power of conventional wisdom and the problems faced by an idea that is deemed to be "guilty." The case has to do with two competing belief systems: one is in the realm of religion and the other is in science. The question to be argued is whether religious beliefs are amenable to scientific testing.

A few years ago, one Daysi Fernandez, a mother of three living on welfare, wished to buy tickets in the New York State Lottery and (it is claimed) she asked a young boy of her acquaintance, John Pando, to purchase the tickets on her behalf. John Pando is deeply religious and felt that if he prayed to St. Eleggua for help, Mrs. Fernandez would have a much better chance of winning the lottery. According to John's story, Mrs. Fernandez gave him four dollars, with which he bought tickets for the lottery in her name. He says she agreed to give him half the proceeds if one of the tickets won. Then he prayed.

One of the lottery tickets was chosen and Mrs. Fernandez won $2,877,203.30. She did not split with John Pando and he took the matter to court to try to recover his half. Mrs. Fernandez argued that the alleged agreement was illegal or unenforceable for a number of reasons, including the fact that John Pando was under 18 years of age.

On October 19, 1984, Judge Edward J. Greenfield of the Supreme Court of New York County ruled on the case. Whereas he found in favor of John Pando on most of the issues, including the problem of age, Judge Greenfield ruled that John had no case because it was *impossible to prove in a court of law that "faith and prayers brought about a miracle and caused defendant to win."*

I agree that John Pando did not prove that St. Eleggua had influenced the lottery, and thus I accept the judge's decision. But I do not accept the reasons for that decision as described in the written decision.

Judge Greenfield's lengthy decision makes interesting reading because he is attempting to show that religion is not a science. Let me quote a few passages:

> Who is going to provide proof that his prayers were efficacious, and that the saint caused the numbers to win? It is not a sufficient answer that he prayed, and that one of the tickets he filled out was the winner. That would leave a gap in the proof, which must demonstrate not merely that winning followed prayer, but that . . . prayer was the causative factor in winning.

> Under Roman Law, there was acceptance of divine testimonies, omens, auguries or oracles and the power of dreams. . . . But in those days, the function of the secular and the ecclesiastical courts was not sharply separated . . . In this more workaday and pragmatic era, shaped by tragic experience, the chasm between the temporal and the spiritual world has become unbridgeable. Theology is to be protected against the law, just as the law is to be protected from theology.

> The condition was not that the numbers chosen would win, but that the saint was to make the numbers win. Establishing that this occurred is not susceptible to forensic proof. It calls for

matters which transcend proof—the existence of saints, the power of prayer, and divine intervention in temporal affairs.

These passages do not mention science explicitly, but as will become clear, the judge was thinking "science" in nearly all of his references to "the law." The text of his decision is filled out with citations to St. Paul, Wordsworth, St. Augustine, and the usual array of legal precedents. But the message is clear: the actions of saints are not amenable to testing. Judge Greenfield did emphasize, however, that:

This court has no desire to denigrate the power of prayer, matters of spirit, or the workings of the hand of God . . .

To me, the real climax of Judge Greenfield's decision comes when he compares the lottery case to hypothetical examples of rainmaking—by both Indian rainmakers and applied meteorologists—because it is here that he hits the problem of belief systems head on. He writes:

If a rainmaker exacts a promise from a group of farmers to be paid if he makes it rain, he can collect if . . . he seeded supercooled clouds with silver iodide and an expert testifies that was the cause of the rain. On the other hand, if the rainmaker performs chants and dances and incantations and it rains within 24 hours, he cannot demonstrate by accepted judicial modes of proof that his acts caused the desired event.

There are two obvious problems with this. First, the judge is assuming, absolutely and *a priori,* that the efficacy of chants and dances cannot be demonstrated. He is not saying just that the power of chants and dances *has not* been demonstrated; rather, he is saying that the power *can not* be tested. This is ridiculous! It would be a straightforward matter to test this by standard methods of experimental biology, as follows.

Hire one or more established Indian rainmakers and run the experiment. The rainmakers could be given designated areas; other areas, without rainmakers, would be used as controls. A substantial number of test areas with and without rainmakers would be needed so that chance differences in microclimate would not affect the statistical results. Rainfall would be monitored in all areas and standard statistical tests used to see if the areas with rainmakers had significantly more rainfall than those without rainmakers. The experimental procedure would have to be designed carefully, but many areas of science do well at this and the problem at hand is relatively simple.

My point is that the efficacy of chants and dances is just as testable as any other scientific proposition—such as whether a particular medical treatment works or not. By arguing that chants and dances cannot be tested, I think Judge Greenfield is really saying that he, speaking for the law and society, does not believe chants work, and therefore chants are not worth the effort of testing. This sounds very much like a commitment to a belief system—something that is anathema to most scientists (and probably judges).

You may say that hundreds of years of human experience have shown that things like chants and dances do not increase rainfall or have any other effects on the natural world. And I say in response that I would like to see the statistics. It is just not fair to rule out a proposition by fiat or anecdote.

The second problem with Judge Greenfield's rainmaking is that seeding clouds with silver iodide may or may not work! It is a technique that has been tried for years by applied meteorologists, and a number of controlled experiments far more elaborate than the one I suggested have been performed. The results are equivocal. Some studies show positive results, but in most cases the outcomes are impossible to distinguish from chance. One major study showed what appeared to be a *decrease* in rainfall as a consequence of cloud seeding. The question has been debated for many years in meteorology. In a court case of the kind Judge Greenfield described, it would be a fairly easy matter to line up "experts" on each side. I am not close enough to that field to go into detail but my impression is that there is a spectrum from "believers" to "non-believers" on the question of cloud seeding. Where does a belief system end and science start?

Let me say that I have no reason to believe that chants and dances produce ràin, nor do I think saints influence the New York State Lottery. My problem is with the unsupported claim that such hypotheses are not testable hypotheses within a conventional framework of science. I would argue that they are indeed testable. We say they are not because we have already decided that the hypotheses are wrong. The conventional wisdom is so strong that the proponents of the religious side of the argument are not even allowed to have their hypotheses tested.

THE LUNATIC FRINGE

Science is surrounded by a lunatic fringe which may strangely affect the progress of science itself. Perhaps once a month, I receive a manuscript or privately printed volume from someone sporting a new theory or paradigm in search of a hearing. There are many such in evolution, Earth history, astronomy, and, I presume, all fields of science. The authors are typically untrained in conventional science or their training is in a distant field of science. There are physicians, engineers, accountants, and a surprising number of independently wealthy business people. My impression is that they are nearly all honest and genuine people. Typically, they have worked for twenty or thirty years to marshal evidence for their particular idea. The tragedy is that virtually all the treatises they produce are a hopeless jumble of misinformation and logical error. One must be careful about such condemnations, of course, because it is always possible that the lunatic fringe is correct and everyone else is wrong. But the cases I am talking about are pretty obvious.

Occasionally, one of these unconventional ideas takes off, at least for lay audiences. Erich Von Daniken's books about visitors from outer space (*Chariots of the Gods* and others), for example. Even more rarely, a person of this genre turns up in a conventional scholarly setting, causing all sorts of problems for mainline science. An example is Dr. Roy P. Mackal. He is what he calls a "cryptozoologist" and has devoted vast amounts of time and energy in recent years to searching for living animals long thought to be extinct. In particular, he has been on several expeditions to Central Africa in search of living dinosaurs—for whose existence he feels there is evidence in local folklore. The press normally describes Dr. Mackal as a "University of Chicago biologist." This gives him some credibility. In fact, he is a biologist and he is employed at Chicago, but as Safety and Energy Coordinator.

Mackal's views on the possibility of living dinosaurs are seen as totally incredible by all biologists I know, and one can marshal impressive arguments to show that the probability of surviving dinosaurs is vanishingly small. But because of the impossibility of proving a negative, one cannot say that there is no chance that dinosaurs still live in remote areas in Africa. I am comfortable with the notion that Mackal is wrong but I cannot prove it.

Another example is Linus Pauling. As a Nobel chemist with an almost unequaled record of achievement, Pauling is one of the world's truly great scientists. But to some observers, his research on vitamin C puts him in the lunatic fringe. Attempts have been made to suppress some of Pauling's research papers, and some of his colleagues in the National Academy see him as an embarrassment. On the other hand, the vitamin C work constitutes a foray into nutrition and medicine where there exist strong paradigms that do not readily accept discordant ideas from a chemist. This may be a case of a brilliant man seeing more clearly. Or Pauling may simply have goofed because unfamiliar with the details of fields outside his normal domain.

The lunatic fringe is an ever-present factor in science. It is loathed and feared, and, inevitably, science protects itself. It is extremely difficult for a really radical idea to get a hearing, much less a fair hearing. And if the originator of the radical idea does not have normal credentials, getting a hearing can be virtually impossible.

To some geologists and paleontologists, Luis and Walter Alvarez, Rich Muller, Jack Sepkoski, and I are probably part of the lunatic fringe.

WEGENER AND CONTINENTAL DRIFT

One should not try to discuss the subject of belief systems in science without at least mentioning Alfred Wegener and his theories of continental drift. Wegener, a highly respected German meteorologist and climatologist, was active during much of the first third of this century. Among other things, he did major exploratory work on the Greenland ice cap (and died there). But more than anything else, he was a synthesizer of world geology. He is best known for his theory of continental drift: the idea that continents are in motion that world geography has changed dramatically over geologic time.

Wegener's evidence was partly geographic, such as the fit of eastern margin of South America into the western margin of Africa, and partly geologic and paleontologic. Working with climatic evidence from the distribution of fossil plants, for example, he postulated not only past movements of continents but also movements of the poles of the Earth. Although his ideas had many adherents, a strong consensus developed against Wegener's hypothesis. The consensus long domi-

nated geologic thought except in a few small pockets—
mostly in South Africa, India, and Australia—where some
of his original evidence had been strongest.

For an American geology student, even as late as the early
1960s, Wegener was anathema. He was not assigned to the
lunatic fringe—his other contributions spared him that—but
he was close. He was mentioned in textbooks and classes only
for comic relief and to provide a bit of the history of geology.
Papers supporting drifting continents rarely got through
peer review—and often were rejected by editors without review.

The strong conventional wisdom of the time was that
continents and ocean basins are permanent, fixed; there was
no "compelling reason" to challenge this. Also, there was no
known physical mechanism for continental movement.

As most readers will know, this is all changed. We now
have drifting continents as part of the larger paradigm of
plate tectonics. The Himalayan mountains were formed
when India, which had been drifting quite rapidly northward
across the Indian Ocean, slammed into southern Asia and
crumpled that continent. And so on. This paradigm shift has
been hailed a major revolution in the history of science.

How did the shift in thinking occur and how long did it
take? This is a little difficult to reconstruct, because memo-
ries are short and untrustworthy, although a number of good
works in the history of science have been devoted to the
problem (especially work by William Glen). The first break
came in the late 1950s with the discovery of a new kind of
data: records of past magnetic fields (already mentioned in
another context in Chapter ii). In spite of the compelling
nature of the new magnetic data, about ten years of furious
and often acrimonious debate were needed for the scientific
community to accept that continents move. By now, nearly
all the original holdouts are either convinced or dead and we
have a new strong paradigm.

Comparing the continental-drift literature with that of the Nemesis Affair, one can easily see a similarity in tone and method of argument. Continental drift was guilty until proven innocent.

BELIEF SYSTEMS AND THE NEMESIS AFFAIR

My question is: To what extent have the events described in this book been influenced by preconceived belief systems? Have the scientists who study Nemesis been influenced by their own biases? If so, is the effect important or trivial? In developing the argument, I do not wish to accuse my colleagues of lacking objectivity. This will make my job a bit difficult. One way to do it is to talk primarily about my own reactions to situations as they arose.

I have already described my early reactions to Al Fischer's 32-million-year periodicity in extinction. I found just about everything wrong with his analysis, from his choice of data to his interpretations. Later, of course, I was in total agreement with his conclusions, after Jack and I had found our own evidence for periodicity. It makes no difference in this context whether periodic extinction is real or not. What is important is the fact that nothing in Al's analysis had changed between the beginning and the end of the sequence: his work was in hard print from 1977 on. One could argue that his analysis was flawed and that I changed my views only about the conclusions, not the way he got there.

A better example may be my review for *Science* of the early version of the Alvarez *et al.* paper in 1980. I have already described the essentials in Chapter 4. But as I look back over that review, I see that I was challenging things in the Alvarez paper that I would now call minor misdemeanors, if that. In spite of my preconditioning to be receptive to impacts as a cause of extinction, I was finding fault in what was pretty surely a classic reactionary mode. If I were to receive the same paper now, describing perhaps an iridium anomaly at some other point in the geologic record, I am reasonably sure I would accept most of the things I found fault with in 1980, because I am now a "believer" in large-body impacts and their iridium signature.

I have to ask the reader to take the foregoing on faith. But let me strengthen it with an anecdote. Sometime late in 1980, I was at a cocktail party in the home of a paleontologist at the University of Texas at Austin. Iridium and the extinction of the dinosaurs were standard cocktail conversation at that time, and I was in a small group talking about just that. The flavor of most comments tended to be negative. I was commenting on the Alvarez paper and was on the verge of saying: "I don't know where Walter Alvarez got his training but he certainly is a lousy geologist." But before I could get the sentence started, a most respected structural geologist by the name of John Maxwell said, "And Walter Alvarez was the best student I ever had at Princeton." Maxwell had recently moved from Princeton to Texas, and then I knew where Walter had gotten his training.

My point is that the remark on my lips was pure prejudice. My sole contact with Walter Alvarez had been in reading the one manuscript, and that was devoted largely to the physics of his father and the chemistry of his two other colleagues. I was smearing a research result I did not like. This is not a rare event in science.

An important control in the Nemesis Affair has been, and continues to be, the Lyellian philosophy. Remember from Chapter 2 what Charles Lyell said:

> . . . we are not authorized in the infancy of our science, to recur to extraordinary agents. We shall adhere to this plan . . . because . . . history informs us that this method has always put geologists on the road that leads to truth.

This sounds very much like the revealed truth of the fundamentalist. And the catechism is still very much with us, although this is sometimes difficult to prove because the wording changes. Like the *New York Times* editorial cited in Chapter 10, which ended with:

> Astronomers should leave to astrologers the task of seeking the cause of earthly events in the stars.

This says, as Judge Greenfield said, that extraterrestrial causes for events on Earth can be ruled out automatically—without doing any research. I suspect that the day will come, perhaps rather soon, when we will look back on this period and wonder how we ever thought the Earth could have been unaffected by all those things whirling around above our heads.

Epilogue

I STARTED this book with two purposes. One was to describe the events that led up to the Nemesis theory and to explore the scientific problem of mass extinction as it relates to Nemesis. The other was to say something about how science works: a view from the trenches.

Inevitably, the first purpose is unfulfilled. The story is not yet over. Nemesis has not been found and much controversy rages over whether there is even any need for Nemesis (or Planet X, or what have you). That is, the claim of periodicity in extinction, on which Nemesis is based, is still being debated and new lines of evidence are being sought. Even though the meteorite impact at the K-T boundary is tantamount to fact, the link to biological extinction has not been clinched. And the evidence for impacts at the other extinctions is of variable quality.

New evidence arrives every month—sometimes every week. Much of it is still raw and in need of further work, but that which survives may change the overall picture dramatically, either toward or away from Nemesis. An example is a preliminary study of amino acids at the K-T boundary reported late in 1985 at the annual meeting of the Geological Society of America in Orlando.

Nancy Lee and Jeffrey Bada of the Scripps Institution of Oceanography analyzed for a certain rather obscure organic compound called *alpha-amino isobutyric acid* (AIBA). They chose this compound for the simple reason that it is commonly found in certain kinds of meteorites but occurs only rarely in living organisms. AIBA is thus a potential marker of extraterrestrial material.

Lee and Bada analyzed for AIBA in sedimentary rocks over a range of geologic ages and found it only in samples from the boundary between the Cretaceous and Tertiary— just where it should be from the Alvarez hypothesis of the K-T meteorite impact. Much more needs to be done and the two biogeochemists want to do a lot more checking before writing a full report. There is a possibility that the AIBA they found came from laboratory contamination. They will need to replicate the observations in other K-T boundary samples. But it looks very promising.

The amino-acid study is a good example of a new generation of research on the extinction problem. More and more, specialists are asking questions like: If there was an impact, what specific predictions can be made and how best can they be tested? or: If extinction events have a 26-million-year periodicity, what does this predict about the behavior of other earthbound or cosmic phenomena? The research is rapidly becoming more sharply focused and thereby much more likely to give clean, definitive answers. It is a very exciting time, because the problems are so new that anyone with good training and a cool head might come up with the definitive test or prediction.

What will happen if all the Nemesis and related theories turn out to be wrong? This is, of course, possible. And some of the more speculative theories I have described have a pretty good chance of being wrong.

I don't think much will happen if the theories turn out to be wrong. There will be some red faces, of course, and some gleeful ones as well. I imagine the next meeting of the Society of Vertebrate Paleontology will have quite a celebration. But by and large, the scientific community will absorb the event, because we are all used to the three-steps-forward-and-two-steps-back progress of science. No careers will be ruined and no staff will be reprimanded. In fact, some of the more active participants may receive some professional rewards for having shown the imagination necessary to think up new ideas and stimulate research in new areas. It has been said that one's success as a scientist can be measured more by the number of people he or she puts to work on new problems than by the correctness of specific research results.

Also, regardless of the outcome, so much will have been learned about the facts of extinction that the groundwork will have been laid for an understanding of the history of life that might have eluded us otherwise. So, in my view, only good can come from the Nemesis Affair. But I still hope fervently that Rich Muller finds that Death Star. We can rename it "The Star of Berkeley."

My second purpose in writing this book, the "view from the trenches," has naturally been dominated by the kind of science practiced in paleontology and its newly found bedfellows, geochemistry and astrophysics. Science is an extremely pluralistic endeavor, and the culture varies somewhat from discipline to discipline. Most of the research I have been talking about is done by individuals, or, at the most, three or four people working together. In other fields, high-energy physics for example, the cost of experimental laboratories—plus tradition—means that research is often done by large teams. (I have even seen a physics article where the list of authors takes up more space than the article itself.) But I don't think there are many fundamental differences in the way science is done. All fields have strong conventional paradigms, paradigms constantly under fire. Some of the paradigms are well founded in observation and experiment; others are simply logical constructions that "seem" to work. And every so often, there is a revolution whereby the old paradigm is replaced by a new one. It is a wonderful area of scholarship to be in—as long as you do not take the wisdom of the moment too seriously.

The various fields of science have a well-defined pecking order. There is a range from "soft" science to "hard" science. In the United States, molecular biology, biochemistry, astrophysics, and high-energy physics are now at the hard end. They have great authority in the scientific and general communities, and they are seen as somehow more rigorous and precise than other sciences. At the other end of the spectrum are fields like ecology, paleontology, and behavioral biology. The social sciences could be added at the "soft" end, although many of the sciences higher on the pecking order do not even recognize fields like economics and sociology as sciences.

I think the differences between the hard and soft sciences are more matters of perception—and some good marketing —than fact. Some fields, like economics, are inherently more difficult than others, like physics, because they deal in complex systems with many more unpredictable elements. The softer sciences also tend to have more observational data, and this has the irony of making it more difficult to construct simple, unifying theories. All fields of science have some very soft spots not immediately obvious. I have heard it said in jest that astrophysicists use overhead transparency projectors and Magic Markers rather than prepared lantern slides for lectures because it is easier to change the numbers during the lecture.

Update 1999

Much has happened in the dozen or so years since this book was written, though the pace of discovery and the fever of speculation have slackened somewhat from the heady days of the 1980s. Most important, the smoking gun has been found: a large crater at the Cretaceous-Tertiary (K-T) boundary that validates the Alvarez scenario for the Cretaceous mass extinction. The lack of a suitable crater had long been a natural point of attack for critics of the impact-extinction link. In defending our position, we had a pretty good candidate: the buried Manson Crater in Iowa. The 65-million-year date was perfect but the crater was much too small, a mere 32 kilometers in diameter, to do the job. At one point, Gene Shoemaker attempted to get around the small size by linking Manson with a couple of weakly dated craters in Siberia to suggest that all were produced by the breakup of a single large comet. Though imaginative and plausible, this idea later broke down when it was shown that the geologic date of Manson was seven million years older than the K-T extinction! This was astonishing, because we had all been so confident of the Manson dating.

The new smoking gun is a crater called Chicxulub buried partly under the tip of the Yucatan Peninsula and partly under the Gulf of Mexico. This is a big one—between 200 and 300 kilometers in diameter. Its 65-million-year date seems to be valid and it matches dates of tektites (glassy blobs thrown out by meteorite impact) in Haiti. Being buried under younger rocks, Chicxulub is detectable only by drilling or deep seismic surveying. It was, in fact, discovered by an industry geologist, Glen Penfield, in 1978 while he was prospecting for oil in Mexico. Penfield proposed Chicxulub as the K-T impact point in 1981 but nobody outside the oil industry was listening. Not until the early 1990s did the people working with the Alvarez theory catch on. Subsequent work has shown pretty convincingly that Chicxulub was ground zero for the K-T mass extinction.

The finding and accurate dating of Chicxulub convinced many who had harshly criticized the Alvarez scenario—many, but not all. There remained the problem of the three-meter gap between the youngest dinosaur fossils and the iridium anomaly in Montana (Chapter 5) and other claims that the extinctions actually preceded the meteorite impact. But the continuing controversy produced a wave of intensive fossil collecting in the critical intervals. Groups of volunteers led by Peter Sheehan of the Milwaukee Public Museum, searching for dinosaur fossils in eastern Montana, narrowed the gap to almost zero. Other fossil groups that had been thought to have died out gradually were collected again. A notable example is Peter Ward's work up and down the Biscay Coast of Spain and France. We knew that the ammonites, a large and important group of fossils, had died out at the end of the Cretaceous but this had been thought to have happened gradually over several million years. Ward found ammonites at virtually full diversity right up to impact time. This and other cases have convinced most paleontologists that the K-T mass extinction was indeed a sudden, catastrophic event.

Some critics of the link between meteorite impact and mass extinction nevertheless remain unconvinced. One is John Briggs, a distinguished marine biologist and professor emeritus at the University of South Florida. He argues that paleontologists have vastly exaggerated the intensity and suddenness of past extinction events and that the idea of sudden annihilations by impact (or volcanism, for that matter) is an unfortunate return to discredited theories of catastrophism of the nineteenth century. Briggs concludes (see reference below) that "The only event that deserves the title of mass extinction is that which is going on right now." I don't think Briggs is right but he is a good biologist and we have to listen. I do think that the several mass extinctions of the geologic past (see p. 37) were real pulses of rapid killing of large numbers of species. The fossil record is quite clear on this. Whether these events were more or less profound than extinctions occurring today is debatable—and probably unknowable—because the time scales and our values are different.

Despite John Briggs and a few others, consensus runs strongly toward the view that mass extinctions are sudden and that the Cretaceous event followed the impact of a large asteroid or comet. But what of the other mass extinctions in the fossil record? In Chapter 6, I discussed five other possible impact-extinction pairs. These and a couple others are still being investigated, with continuing wrangles over uncertainties of geologic dating and quality of evidence for this or that impact. The greatest die-off in history was the event at the end of the Permian, 250 million years ago. I referred (pp. 99–100) to a Chinese report of an iridium anomaly but noted that it had not been confirmed. Many analyses have been done since and it is clear that there is no authentic anomaly at or near the Permian extinction. This does not rule out an impact, of course, because the impactor could have been an object without significant iridium (e.g., a clean comet), but this, along with the failure to find shocked quartz, a crater of the right age, or other indication of impact makes the Permian a decidedly poor candidate for extinction-by-impact!

For the Permian extinction, an entirely different though still catastrophic explanation is emerging. Our knowledge of that extinction is far better than it was a decade ago thanks to intensive fieldwork in China. The extinction now appears to have been much more sudden than previously thought—lasting perhaps as little as 10,000 years and certainly less than a million years. Also, recent radiometric dating of the volcanic eruptions that produced the vast Siberian Traps (basalt flows) to the north shows that the timing of the volcanism is right on the best dates for the mass extinction. The Siberian Traps are the largest such deposits ever formed on land. Perhaps we have here another smoking gun. If so, we are bound to ask whether the Siberian volcanism was in itself triggered by a meteorite impact. This sort of link was suggested by some, you will remember (pp. 93–94), to explain the similarity in ages of the Deccan Traps in India and the K-T extinction.

Regardless of how these and other cases turn out, the possibility of annihilation of large numbers of species by rare catastrophic events, be they impacting asteroids or earthbound volcanism, is on our research agendas. These causes must now be considered plausible candidates. Gone are the days when everyone "knew" that rocks do not fall out of the sky and that all natural processes are slow and all change gradual. Our view of geologic history is, I think, permanently changed.

What of the 26-million-year periodicity and Nemesis? No companion star or tenth planet has been found although the skies have been diligently searched, especially by Rich Muller at Berkeley. Is this because we are looking for a needle in a haystack or because the whole idea of extinction being driven by periodic events in the solar system or galaxy is wrong? I don't know. At the heart of the problem is deciding whether or not extinction is truly periodic. Is the interpretation of the fossil record that Jack Sepkoski and I put forward correct? If not, the Nemesis star and other explanations offered by astronomers are meaningless.

Since our original paper in 1984, Jack has increased the size of the extinction database tenfold. He has also refined the time scale so that we have more points in the time series and thus pulses of extinction can be defined more precisely. The results are much cleaner and I doubt we would have had as much argument over our statistics in the 1980s had we waited for the better data. Nevertheless, there has been and continues to be debate over the validity of our 26-million-year periodicity. In 1991, I undertook a review of the problem (see reference below), looking at the reanalyses of our data by other scientists. Between 1984 and 1987, thirteen papers were published that reanalyzed Jack's extinction data (most based on the original database). Of these, eight found no convincing periodicity and five found our periodicity to be statistically acceptable, with some variation in cycle length in a couple of cases. Does this mean that periodicity loses by a score of 8 to 5? Certainly not! It does mean, I think, that more data are needed. Ideally, we should have a totally independent kind of data. For example, a complete chronology of impact craters on the Moon would provide the ideal means of assessing periodicity in impact rates. If such a chronology also showed a 26-million-year periodicity, the extinction case would be strengthened immeasurably. Alas, the Apollo missions did not date enough lunar craters. Several indirect means of assessing the timing of lunar impacts are being explored. So, more later, perhaps.

For the time being, however, the periodicity question is firmly planted on the back burner. Jack and I and some others are still optimistic but the broader consensus in the scientific community is that extinctions are probably not periodic. For one thing, the statistical analysis has been questioned; for another, no convincing mechanism (Nemesis, etc.) has been established. We'll see. I am heartened by the fact that plenty of astronomers, geologists, and paleontologists remember the periodicity proposal and will be quick to revive it should new data come along to support it.

Having convinced you, I hope, that the great mass extinctions in the history of life were often (or usually) triggered by rare, physical events, let me warn you that a very different theory is emerging. It is an offshoot of chaos mathematics and part of the new science of Complexity. The central claim put forward by students of complexity is that species in a plant or animal community compete with one another in a complex system of feedbacks that, in turn, determine the success or failure of all species in the community. Occasionally, a whole system may collapse and the species die out as a consequence of the internal dynamics of the community itself. Thus, simultaneous extinction of interacting species can occur without any special external stimulus. The book by Per Bak cited below summarizes this approach in a readable form. Michael Crichton is also taken by the idea: In *The Lost World* (1995), a main character, Ian Malcolm, is a strong advocate of mass extinction as the inevitable result of complex interacting systems. In the book, theories of meteorite impact and 26-million-year periodicities are relegated to the realm of "frivolous and irrelevant speculation."

The complex-system approach sounds reasonable: it is a way of thinking about ecology familiar to most schoolchildren. And collapse of ecosystems can occasionally be observed in very simple systems involving, for example, overgrazing of fragile plant communities. But watch out! Processes that work in small systems may not be transferable to larger ones. "Complexologists" have developed elaborate computer simulation models to explore systems with many species in heterogeneous environments in an attempt to test the extrapolation from small to large. Computer simulation has been necessary because the many interactions are too complex (and often have too many random elements) for precise prediction of outcomes and because mass extinction due to collapse of internal dynamics has not been observed in the real world. Several of the simulation efforts have produced patterns reminiscent of mass extinction. This has led to the claim that events like the K-T mass extinction can be explained without recourse to external causes. Is this credible? I don't think so. The main problem is that events like the K-T involve extinctions of species that do not interact with one another because they live in totally different environments in different parts of the world—for example, dinosaurs in Montana and ammonites in Antarctic seas. I realize that conventional wisdom these days sees the Earth's biosphere as one great ecosystem but the level of interdependency required by this theory is to me, at least, beyond reason. To be sure, I did not like either the Luis Alvarez idea of rocks falling out of the sky or Al Fischer's periodicity. We will have to see how the complexity idea plays out and, in particular, whether its proponents can come up with hard evidence.

I provide below a few references for further reading. Please keep in mind that I do not agree with the views of Per Bak or John Briggs. But read them and decide for yourself. Above all, please enjoy one of the greatest debates in modern science!

SOURCES AND FURTHER READING:

Alvarez, Walter. 1997. *T. rex and the Crater of Doom.* Princeton University Press.

Bak, Per. 1996. *How Nature Works: The Science of Self-Organized Criticality.* Springer-Verlag.

Briggs, John C. 1998. "Biotic Replacements—Extinction or Clade Interaction?" *BioScience,* vol. 48, pp. 389–395.

Glen, William. 1994. *The Mass Extinction Debates: How Science Works in a Crisis.* Stanford University Press.

Gore, Rick. 1989. "Extinctions: What Caused the Earth's Great Dyings?" *National Geographic.* pp. 662–699.

Muller, Richard A. 1988. *Nemesis: The Death Star.* Weidenfeld and Nicolson.

Raup, David M. 1991. *Extinction: Bad Genes or Bad Luck?* W. W. Norton.

Raup, David M. 1991. "Periodicity of Extinction: a Review." *In Controversies in Modern Geology* (D. W. Muller, J. A. McKenzie, and W. Weisett, eds.). Academic Press. pp. 193–208.

Sheehan, Peter M., et al. 1991. "Sudden Extinction of the Dinosaurs: Latest Cretaceous, Upper Great Plains, USA." *Science,* vol. 254, pp. 835–839.

Index to the first edition